Heavy-Tailed Distributions in Disaster Analysis

Advances in Natural and Technological Hazards Research

Volume 30

For other titles published in this series go to
http://www.springer.com/series/6362

V. Pisarenko · M. Rodkin

Heavy-Tailed Distributions in Disaster Analysis

Springer

Dr. V. Pisarenko
International Institute of Earthquake
Prediction
Theory and Mathematical Geophysics
Russian Academy of Sciences
Moscow, Russia
pisarenko@yasenevo.ru

Dr. M. Rodkin
International Institute of Earthquake
Prediction
Theory and Mathematical Geophysics
Russian Academy of Sciences
Moscow, Russia
rodkin@mitp.ru

ISBN 978-90-481-9170-3 e-ISBN 978-90-481-9171-0
DOI 10.1007/978-90-481-9171-0
Springer Dordrecht Heidelberg London New York

Library of Congress Control Number: 2010931864

Printed on acid-free paper

Springer is part of Springer Science+Business Media (www.springer.com)

Introduction

The end of the twenty-first century saw a dramatically increased interest in safety and reduction of losses from natural and manmade disasters. The cause of this was the combination of a strongly felt social need and the appearance of new theoretical approaches that have significantly progressed in this important interdisciplinary area of study. We wish to emphasize that the increased social need was caused, not only by the (conventionally stressed) growth of the losses due to natural and manmade disasters, but also by the growing interrelationships between different regions of the world. To take one example, the seismic disasters which occurred in China in 1920 and 1927 and entailed the loss of about 200,000 lives each, went almost unnoticed in Europe and in the rest of the world. The similar-sized seismic disasters of 1976 and 2004 which occurred in China again and in Sumatra reverberated throughout the world and stimulated the development of geophysical research and major national and international measures taken to reduce the losses due to possible reoccurrence of such events.

The energy of even a moderate-sized natural disaster and the associated losses is very large. The number of casualties and the loss due to significant natural disasters are comparable with those resulting from regional military conflicts. The energy of natural disasters still exceeds the energy potential of mankind. For example, the yield of the largest nuclear bomb (58 megatons of TNT) detonated in 1961 was 24 $\times$ 10^{23} ergs, while the energy of a major hurricane (recurring at a rate of about two events per year) is estimated to be about 3 $\times$ 10^{25} ergs. The energy of the elastic waves excited by an earthquake with an average rate of occurrence once a year is about 6 $\times$ 10^{23} ergs, while its total energy is about two orders greater, and the total energy of rare great earthquakes is approximately two orders greater still.

The study of disaster statistics and disaster occurrence (here and below, we mean natural and manmade disasters, not catastrophes in a strictly mathematical sense) is a complicated interdisciplinary field involving an intimate interplay of new theoretical results from several branches of mathematics, physics, and computer science, as well as some important applied problems, including socio-economic ones. It is usually thought [VVM] that this research area "is a connecting link between natural, technical, and social sciences". This interdisciplinary character of the area is reflected in the present monograph, which discusses both the limit theorems in mathematical statistics and a possibility of practical realization of the sustainable development of mankind.

There has been little progress so far in the study of disasters (catastrophes) which are sudden and rare processes, and which therefore are little amenable to analysis. The progress that can be seen today in that field crucially depends on the new theoretical approaches that have been developed in several areas of physics and mathematics during the later half of the twentieth century and on the advanced systems now available for environmental monitoring. Understanding the nature of catastrophes essentially relies on new theoretical approaches such as the mathematical theory of catastrophes by Thom and the theory of dissipative structures due to I. Prigogine; other important contributions include P. Bak's concept of self-organized criticality, M.A. Sadovsky's concepts of hierarchical structure and inner activity of the geophysical medium, as well as several other new approaches and conceptual innovations. The statistical studies in the mode of occurrence of natural disasters largely rely on fundamental results in the statistics of rare events derived in the twentieth century. In this connection one can mention R. Fisher, D. Tippet, R. von Mises, E. Gumbel, B.V. Gnedenko, J. Pickands, K. Pickands, and Ya. Galambos.

In relation to natural disasters, it is not so much the fact that the importance of this problem for mankind had been realized during the last third of the twentieth century (the myths one encounters in ancient civilizations show that the problem of disasters has always been urgent), but the realization of mankind's possessing the necessary theoretical and practical tools for effective studies of natural disasters with consequent effective, major practical measures taken to reduce the respective losses. The realization of this situation found its expression in the International Decade for Natural Disaster Reduction adopted by the UN General Assembly in 1989 and in numerous national programs for loss reduction.

All the above factors combine to facilitate considerable progress in natural disaster research. The accumulation of factual material relating to various kinds of natural disasters and the use of advanced recording techniques have expanded possibilities for the analysis of empirical distributions of disaster characteristics. The necessary terminological basis had been developed by N.V. Shebalin and his associates in terms of geophysical magnitude, intensity, and disaster category [She, RS1, RS2].

However, despite the considerable progress reached, the situation with the study of disasters is still far from desirable. It was noted in a review [VMO] that sufficiently complete catalogs of events are still not available for many types of disaster, and the methodological and even terminological bases of research need further development and unification. Not also that the methods of theory of catastrophes and corresponding mathematical approaches are of limited applicability in the very majority of natural disasters because the corresponding potential function is unknown.

The present monograph summarizes our long-continued work in the field of disaster statistics and related questions. We provide a brief description of the terminology and of several modeling approaches and use a broad range of empirical data on a variety of natural disasters. Graded by the amount of factual material, the focus is on seismicity and earthquake loss data, less attention being given to hurricane observations. We also use data on the maximum discharge of rivers, volcanic eruptions, sea-level surges, climatic fluctuations, and manmade disasters.

The main focus is on the occurrence of disasters that can be described by power-law distributions with heavy tails. These disasters typically occur in a very broad range of scales, the rare greatest events being capable of causing losses comparable with the total losses due to all the other (smaller) disasters of this type. The disasters of such type are the most sudden and entail great losses of human life. It is this type of disaster which is meant by the word "disaster" or "catastrophe" in mass media and in mass consciousness. The mode of occurrence and statistics of these disasters are considered for seismic disasters, the information on these being the most complete compared with the others. The monograph contains several new results in the statistics of rare large events. One of the most important results is the conclusion about instability to which the "maximum possible earthquake" parameter is subject, this parameter being frequently used in seismic risk assessment. We suggest alternative and robust ways to parameterize the tail of the earthquake frequency–magnitude relation.

Several results derived in analysis of earthquake loss data seem to hold a universally human interest. For example, the analysis of earthquake losses suggests a revision of the well-known pessimistic forecast of rapid increase of losses from natural disasters; according to this forecast, the increase in the economic potential will as early as in the mid-twenty-first century be entirely consumed by increased losses from natural disasters. We show that the increase in the total number of reported earthquakes is caused by enhanced detection capabilities resulting in the reporting of smaller events rather than by a real increase in the vulnerability of society or increased seismic activity. As to the nonlinear time-dependent growth of total earthquake losses, this effect can be explained (at least the bulk of it) by the peculiar power-law distribution of losses and thus is hardly related to a deterioration of the geoecological environment.

Comparison of the earthquake losses in regions with different levels of economic development suggests a decrease of death toll in economically developed countries. This tendency will extend in the course of time to the third world countries. While the trend of increasing absolute material losses will continue, it is to be expected that the normalized losses (divided by per capita income) will be comparatively stable and even decreasing.

On the whole, the analysis of the relationship between earthquake losses and socio-economic conditions suggests that some sort of equilibrium exists, and that the statistics of losses is compatible with sustainable development.

The monograph has the following structure. The first chapter provides a general overview of the problem, quotes data on different kinds of natural disasters, and gives a classification of these. The second chapter discusses conditions that favor the occurrence of distribution laws typical of disaster magnitude values and gives a consistent description of disasters in terms of distributions and in terms of the mode of occurrence for individual events. The third chapter considers methods in use for a nonparametric description of disasters; these methods are of considerable practical interest in those cases where the actual distribution is either unknown or debatable. The fourth chapter discusses the nonlinear growth over time for characteristic values of the total effect caused by events that obey a heavy-tailed distribution. In the fifth chapter we investigate the distribution of earthquake seismic moment and define the notion of the maximum characteristic earthquake. The sixth chapter is concerned with the distribution of rare, extremely large earthquakes, demonstrating that the maximum possible earthquake concept which is widely used in seismic zonation suffers from instability. The alternative robust approach to parameterization of the tail of the earthquake frequency-magnitude relation is suggested and applied to several cases. The seventh chapter discusses relationships between earthquake losses and the socio-economic conditions, also developing a forecast for characteristic loss values.

This monograph aims primarily at specialists in the field of seismology and seismic risk, but could also be useful for those interested in other kinds of natural and manmade disasters. The basic statistical results (which are largely due to the present authors) are set forth both at the professional and the popular level, thus making them accessible to readers with no special mathematical background. This makes the monograph useful for workers in regional and federal governing bodies, as well as for a broad class of readers interested in the problems of natural disasters and in their effect on the development of mankind. We sought to facilitate the understanding of the problems under discussion (also for nonprofessionals in geophysics and mathematical aspects of risk assessment) by summarizing the results of each chapter at the end of that chapter and outlining the relationships between these and other sections of the book.

This book is a revised and enlarged version of the previous Russian edition of the monograph by the same authors “Heavy-Tailed Distributions: Applications to Disaster Analysis”, Computational Seismology, Issue 38, Moscow, GEOS, 2007. The authors are grateful to Prof. A.A.Soloviev (Executive Editor of the Issue) for permission to use the materials of the Russian Edition. Some authors’ materials published earlier in a number of papers were used also. These materials are reproduced by permission of the “Russian Journal of Earth sciences” and “Fizika Zemli (Izvestiya, Physics of the Solid Earth)”. The authors are very grateful to A.L. Petrosyan for his valuable help in making the English more readable.

We are grateful to the late Nikolai Vissarionovich Shebalin, who was our friend and colleague and provided general guidance of, and himself took an active part in the cycle of studies which have led to the writing of this book. The consideration of several important problems would have been impossible without the initiative and participation of G.S. Golitsyn (Institute of Physics of the Atmosphere, Russian Academy of Sciences, RAS), M.V. Bolgov (Institute of Water Problems RAS), I.V. Kuznetsov (International Institute of Earthquake Prediction Theory and Mathematical Geophysics RAS), A.V. Lander (International Institute of Earthquake Prediction Theory and Mathematical Geophysics RAS), A.A. Lyubushin (Institute of Physics of the Earth RAS), D. Sornette and A. Sornette (Federal Institute of

Technology, Switzerland), Yu.A. Shishkov (Institute of Oceanology RAS), and M. I. Yaroshevich (NPO Taifun Rosgidromet). To all these we are sincerely grateful. We also thank O.D. Voevoda, N.P. Grudeva, T.N. Tseplinskaya, L.S. Kadlubovich, O.A. Matsievskaya, and T.A. Rukavishnikova for the aid they gave during the writing of this book and for valuable advice which has improved the style.

Our disaster studies were supported by the Russian Foundation for Basic Research (grants 93-05-09193, 95-05-14815, 97-05-64583, 99-05-64924, 02-05-64379, 05-05-65276, 09-05-01039, 09-05-92655), by Program 16 of the RAS Presidium "Climate and Environment Changes: Natural Disasters", and by the Program of State Support of Leading Scientific Schools of the Russian Federation, projects no. NSh-5009.2006.5.

Contents

Chapter 1
Distributions of Characteristics of Natural Disasters: Data and Classification

Any useful classification contains three to six categories.

From scientists' folklore

1.1 The Problem of Parameterization and Classification of Disasters

This chapter is devoted to a review of empirical data on natural disasters, to a discussion of distribution laws for disaster size and to the description of the approaches used for parameterization and classification of disasters. We wish, first, to acquaint the reader with the great variety of statistical and physical characteristics used to describe different kinds of natural and man-induced disasters. As a result of this review, it appears possible to specify the place occupied in the variety of different kinds of disasters and adverse phenomena by a special class of disasters, namely, those described by distributions with "heavy tails". The present book is mainly devoted to results of research in the statistical characteristics of just this kind of disaster. The term "heavy-tailed distribution" is commonly used for a distribution density that decreases at infinity slowly enough, for example, more slowly than any exponent. We will use the term in a narrower sense, namely, for distributions with an infinite mathematical expectation. For such distributions the law of large numbers and the central limit theorem in the theory of probability do not hold, and thus the standard statistical characteristics such as sample mean and variance are inapplicable. The distributions possessing this property – being theoretically with an infinite mean - will be called heavy-tailed distributions in what follows.

The analysis outlined later in this chapter includes not only a review of available data sets on natural disasters (such as the USGS and EM-DAT Data Bases), but also a discussion of the terminology and of the approaches to parameterization and classification of disasters. The need for a discussion of these methodological aspects

V. Pisarenko and M. Rodkin, *Heavy-Tailed Distributions in Disaster Analysis*, Advances in Natural and Technological Hazards Research 30, DOI 10.1007/978-90-481-9171-0_1,

arises from the absence of a commonly used and effective theoretical base and terminology in this interdisciplinary field of research. The situation is aggravated by the use of specific ways for description of disasters and of specific functions in the fitting of empirical data for different kinds of disasters. The great variety of the approaches now in use complicates the comparison of data on different kinds of disasters and the development of a unified system of measures to take for loss reduction following various possible natural and man-induced disasters.

Note that the study of distribution laws for different natural disasters (earthquakes, volcanic eruptions, floods, hurricanes, etc.) is necessary not only for risk assessment, but it is useful also for understanding the physical nature of the underlying processes.

The earthquake statistics is the best-studied case among disasters. The distribution laws for the other kinds of natural disasters are much less known. For most kinds of natural disasters (floods, thunderstorms, etc.) the relevant empirical data on disaster occurrence are traditionally fitted by a specific parametric distribution function. These approaches are often injustified statistically (especially in the range of rare greatest events), and why a particular parametric distribution should be chosen is not quite clear. A formal character of the distribution functions in use and the ambiguity of their physical meaning tend to restrict their applicability for extrapolation into the domain of rare large events.

An alternative to the description of empirical distributions in terms of different historically evolved specific parametric distributions is the use of the "classical" physical distribution laws. These "classical" distributions frequently encountered in physical systems are the Gaussian (normal) distribution, the Boltzmann (exponential), the Pareto power-law distribution, as well as modifications of these. Each of these distributions occurs under some set of conditions that are fairly clear and widespread in natural systems. Consequently, one obvious advantage for the use of these "classical" distributions is the possibility of physical interpretation of the results and various physical analogies of the phenomena. It seems reasonable to assume that, if empirical data for a given kind of disasters can be described by one of these distributions, then the conditions generating disasters of the kind in hand are similar to the conditions for the occurrence of the corresponding "classical" distribution.

In view of the prevalence of conditions that favor the occurrence of classical distributions, they can be considered as the most natural and expected. Accordingly, if a given empirical distribution can be described well enough by one of the classical physical distributions, it is reasonable to consider such parameterization as the more preferable compared with other possible laws, historically evolved or chosen more or less formally. The use of classical distribution laws also inspires some hope for deriving correct results when the parameterization is applied to other similar situations or in the domain of rare events. Considering the aforesaid, preference is given below to the description of empirical distributions by one (or a combination) of classical physical distributions. This approach will prove to be productive enough throughout this book.

The empirical distributions for many kinds of natural disasters have not been sufficiently studied because of shortage of statistical data for corresponding kinds of natural disasters. Insufficient development of the parameterization techniques has also affected the situation. In both these directions considerable progress has been made recently. The use of new recording methods has considerably improved data acquisition. In the methodology, the conceptual scheme developed by [She, RS1, RS2] provides a sound methodological basis for disaster parameterization. It includes explicit definitions of the notions of geophysical magnitude of a disaster, its intensity, and its category.

It would be most natural to use the energy released in a disaster as the magnitude (size) of that disaster (earthquake, hurricane, etc.). However, for most cases the energy of an event cannot be calculated directly, but only an indirect energy-related characteristic can be estimated. The magnitude of disasters is usually calculated from an available, easily derived physical parameter that is connected with energy. An example of such an approach is the classical definition of earthquake magnitude as the logarithm of the maximum amplitude of a certain seismic wave corrected for distance from the epicenter.

The intensity of a disaster characterizes its impact at a given site (upon the objects located nearby: people, buildings and various man-made structures and natural objects). The intensity depends on the magnitude of the disaster, on the distance from the epicenter, and on the site effects. A particular disaster is characterized by a set of intensity values describing the impacts at some points of the epicentral area.

The disaster category is characterized by the damage caused by the disaster. The disaster category depends on several parameters: number of casualties, number of injured, direct and indirect loss of property. Exact values of the loss, as well as a full description of different kinds of the damage are very seldom available, however. Even till very recently, the loss was characterized usually only by the death-toll, and sometimes by direct economic losses.

Before passing to a discussion of the data sets characterizing different kinds of disasters, we wish to note a few general points. Smaller events are often less completely reported in the catalogs of disasters. This reduction can stem from several causes. For the majority of cases we have an incomplete reporting of smaller events. In this connection it is unpromising as a rule to search for the unique distribution function that would be valid in the entire range of magnitude of a given kind of disaster. Possible errors in the number of smaller disasters do not however seriously affect the estimation of losses because of the insignificant input of small disasters into the total loss, despite the large number of these events.

We notice, however, that the reduction is sometimes due to the fact that in the range of small events the fitted law (e.g., the Pareto law) is invalid. Such a situation arose, for example, in predicting the number of smaller deposits on the assumption that their distribution is the self-similar Pareto law [KDS, BU1, RGL].

The next point is connected with the very wide range (several orders of magnitude) of disaster size and disaster loss. Because of this, and also because the empirical loss distributions frequently obey power-law, a log scale was suggested for the classification of disasters [RS1]. Then the category of disaster changes from local (weak) accidents to as far as disasters on a planetary scale. The scheme accepted now in the Ministry of Emergencies of the Russian Federation uses a similar classification system for accidents by severity levels of local, territorial, regional, and federal significance [SAK]. For the same reason many empirical scales of natural disasters (earthquake magnitude and intensity, the Iida tsunami scale, scale of volcanic eruptions) are logarithmic. Other widespread scales of natural effects include the Beaufort wind velocity scale, the Saffir-Simpson scale for hurricanes, the Fujita and Parson scale for tornadoes; all of these are compromises between the linear and logarithmic principles. Detailed descriptions of the basic scales and references to such descriptions are given in [Fu, Bl].

1.2 Empirical Distributions of Physical Parameters of Natural Disasters

The empirical distributions of physical parameters describing different natural and man-induced disasters and adverse phenomena are examined below. The analysis is given in the order of possible data fitting using the normal, exponential, and power-law distributions. All data sets are shown to be suitable material to be fitted by one of these classical distributions or by a combination of these.

The empirical cumulative non-normalized distribution functions of temperature deviations from the mean seasonal values for St.-Petersburg are shown in Fig. 1.1 along with the fitted normal law. As can be seen, these data can be described well

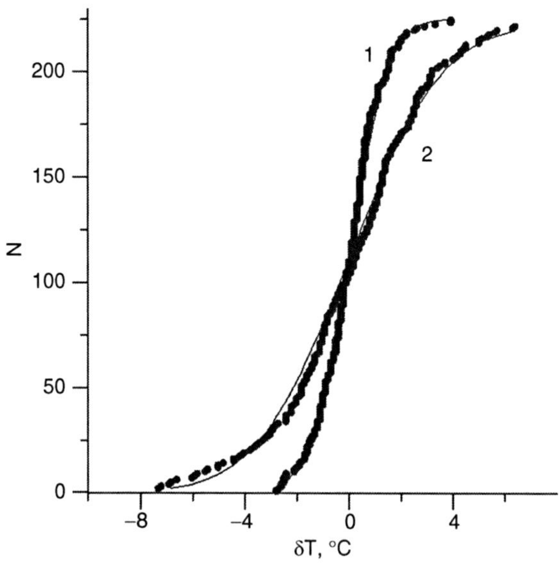

Fig. 1.1 Distribution of δT deviations of seasonal temperature from the means for St.-Petersburg for 1744–1980 (*dots*), data from [Bo], (**1**) summer, (**2**) winter. The *line* gives fitted normal distribution laws

enough by the normal distribution. The chi-square test shows that the data are consistent with the normal distribution at a confidence level greater than 90%. Considering the short time series used (220–240 data points or years), this confidence level appears to be quite acceptable.

The wide prevalence of the normal distribution in application to natural phenomena is due to the conditions required for the occurrence of this law. The normal distribution arises in cases where the parameter of interest depends on a sum of independent factors none of which is prevailing or infinite. Situations such as this are rather common and fulfill the conditions required for the Central Limit Theorem in probability theory on the convergence of distributions of the sum to the normal (Gaussian) law. The distribution function for the normal law is

$$F(x) = \int_{-\infty}^{x} (2\pi\sigma^2)^{-1/2} exp\left[-(t-m)^2/2\sigma^2\right] dt, \tag{1.1}$$

where m, σ are the parameters: mean and variance, respectively.

The probability density of the normal law has the form

$$f(t) = \left(2\pi\sigma^2\right)^{-1/2} exp\left[-(t-m)^2/2\sigma^2\right]. \tag{1.1a}$$

It can be seen that the probability of deviation $\delta\varsigma = (t-m)$ quickly decreases as $\delta\varsigma \to \infty$. The probability of deviation $\delta\varsigma$ exceeding 3σ can often be disregarded in practice (its value is about 0.002).

The next widespread distribution is the Boltzmann exponential law. This distribution law is also typical of catastrophic processes. Its occurrence in physical systems is treated as the distribution of energy values for a set of particles being in thermal equilibrium with a thermostat. Empirical non-normalized complementary distribution functions for sea level variation amplitudes as observed at a number of points along the east coast of Canada for 1965–1975 are presented as examples in Fig. 1.2. Distributions of wind velocity recorded at weather stations are similar in character.

The distribution function for the exponential law is

$$F(x) = \int_{-0}^{x} 1/x_0 \; exp(-t/x_0) dt, \tag{1.2}$$

where the parameter x_0 is the mean. According to a popular physical interpretation, x_0 is equal to the mean energy of particles at temperature T. More exactly, for the case of ideal gas model, we have $x_0 = kT$ where k is the Boltzmann constant. The associated probability density has the form

$$f(t) = 1/x_0 \; exp(-t/x_0). \tag{1.2a}$$

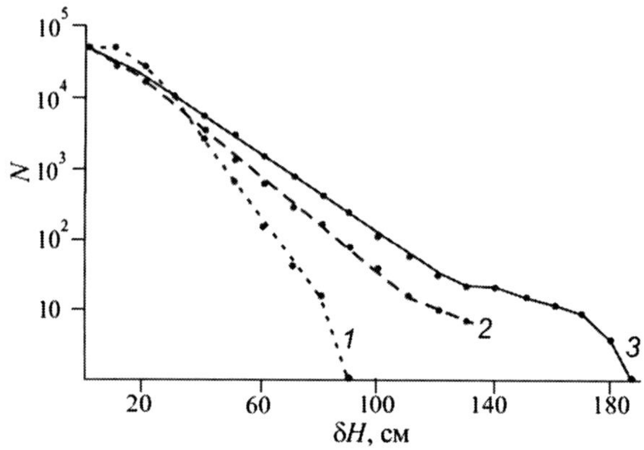

Fig. 1.2 Distribution of the number N of sea level changes with amplitude not less than δH at the east coast of Canada for 1965–1975, data from [EMB]: (**1**) sea tides for the Saint Johns station, (**2**, **3**) tides and ebbs for the Rivier-du-Lop station, respectively

The observed sea level variations and wind velocity distributions [Ku, GD, SY] corroborate the possibility of a good exponential fitting, though the authors do not always use this fit. The distribution parameters based on different data sets vary widely depending on the season and registration conditions (for example, site elevation above the ground surface in measurements of wind velocity).

The case of sea level variations can evidently be interpreted as variations in the potential energy of a system and the empirical distribution can thus be treated as an analogue to the classical Boltzmann distribution. The interpretation is less evident for the case of wind velocity. Naturally, wind velocity itself is not the same thing as energy of weather phenomenon. It seems a plausible hypothesis, however, that wind velocity and thermal energy of the relevant weather process can be connected by a linear regression. Actually, air pressure and wind velocity correlate closely (see below an example of the evolution of hurricanes). At the same time, according to the universal gas law, the pressure change in a gas is proportional to the change in temperature. It is reasonable to suppose that wind velocity is related statistically to temperature variations and consequently to the energy of the associated weather phenomenon. It thus appears that the exponential distribution of wind velocity could be treated as an example of a distribution of the Boltzmann type.

The distributions of wind velocity when measured in hurricanes (also referred to as typhoons and tropical cyclones) and tornadoes can differ from the distribution of wind velocity under ordinary conditions. The statistics of wind velocity for tropical cyclones occurring in the western part of the Pacific Ocean were studied in [GPRY], where the cyclone hazard for a few large cities in Southeast Asia and the Far East was evaluated. According to the United Nations data for 1962–1992, the tropical cyclones cause more than 20% of all loss of life due to natural disasters [O1]. We are going to consider the case of typhoons in more detail.

The data of two catalogs to be referred to in what follows as catalogs 1 and 2 were used. The western part of the Pacific Ocean within the coordinates 0°–55°N and 90°E–160°W was examined. The wind velocities are spatially distributed in a typhoon as follows. A cylindrical region about 30 km across, called "the eye", is located at the center of the typhoon. In this central region the wind velocity and air pressure are the lowest. Around the center there is "an eye wall", which is a circular cylinder whose external radius depends on the intensity of the typhoon and is about 70 km on the average. The wind velocity in this region is the greatest; it can be as high as 80–100 m/s and even more. Farther away toward the periphery of the typhoon the wind velocity decreases, reaching values about 15 m/s at distances of about 500–1,000 km from the center.

Catalog 1 (kindly lent by Ju. A. Shishkov, Institute of Oceanology of the Russian Academy of Sciences, Moscow) contains the coordinates of centers of typhoons and maximum wind velocity values of typhoons in 1950–1988, sampled at intervals of 12 h during the existence of a typhoon. Catalog 2 (from the Hydrometeorological Center of the Russian Federation) contains more detailed information, but on fewer typhoons. The sampling rate of catalog 2 is 6 h. Catalog 2 was used for substantiating the method of risk assessment we proposed. The technique was applied to the data of catalog 1 because this contains more information. The results of this analysis are presented in knots and nautical miles, traditional units in this area of knowledge (see for details [GPRY]).

Here we concentrate on correlations between the characteristics of typhoons and on their interpretation. Firstly, we note a close correlation between the pressure deficit at the center of a tropical cyclone $\delta P(t) = [1{,}010 - \mathrm{P}\,(t)]$ mbars (where P (t) is the lowest pressure at the center of that cyclone) and the squared maximum wind velocity $V^2{}_{max}$. The regression relation of $\delta P(t)$ on $V^2{}_{max}$ is

$$\delta \mathrm{P}(t) = 0.086 V_{max}(t)^2 + 4.31 + \xi, \quad \text{(mbar)}, \tag{1.3}$$

where ξ is a Gaussian variable with zero mean and standard deviation 4.0. The correlation coefficient r between $\delta P(t)$ and $V_{max}(t)^2$ is 0.98. Thus, there is a very close relationship between δP and $V^2{}_{max}$. This statistics is supported by the relation obtained between these characteristics from physical considerations [Ho, Wo]:

$$V_{max} = (b\,\delta \mathrm{P}/\rho)^{1/2}, \tag{1.4}$$

where ρ is air density, and b lies between 1 and 2.5.

The pressure deficit at the center of a typhoon is traditionally considered as characterizing the typhoon "size" [Ho, Go1, Go3]; in the accepted terminology "size" means the magnitude (energy characteristic) of a typhoon. Accordingly, the square of wind velocity $V^2{}_{max}$ can also be treated as typhoon magnitude (energy). We recall that for moderate wind velocities the energy is proportional to the first power of the velocity.

We denote by w_{max} the maximum wind velocity for each typhoon. The distribution of w_{max} for 1,013 typhoons in catalog 1 is shown in Fig. 1.3. In a log plot, this distribution can be described by two lines intersecting at $w_{max} \approx 150$ *knots*.

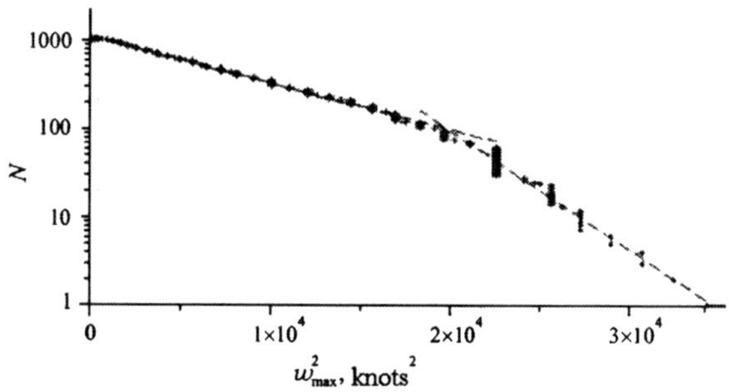

Fig. 1.3 Distribution of number N of typhoons with squared maximum wind velocity not less than w^2_{max} for data base of 1,013 typhoons occurring in 1950–1988; *crosses* denote w_{max} values for different hurricanes, two fitted straight lines are given by *dotted lines*

Thus, roughly speaking, 15% of the greatest values of $w_{max}{}^2$ have an exponential distribution with density $f(x) \sim \exp(-0.0003x)$ for $x > 150^2$, while the remaining 85% have the density $f(x) \sim \exp(-0.0001x)$. A similar exponential distribution is found for the current $V^2{}_{max}(t)$. According to [Go1, GPRY], a typhoon with wind velocity equal to 150 knots has the kinetic energy $E = 3.3 \times 10^{18}\,J$. We thus draw a tentative interpretation of this fact by supposing that the typhoons with kinetic energy E greater and smaller than $E_0 = 3 \times 10^{18}\,J$ differ in structure. However, this change in the character of distribution for moderate and large typhoons is of minor importance, and the distribution of typhoon magnitude (energy) mostly obeys an exponential law. A similar inference for the USA hurricanes can be found at [coastal.er.usgs.gov/hurricane_ forecast/barton4.html, accepted in 2000; Go3].

The exponential law can be used, however, not only to describe the pressure deficit and the squared wind velocity. The variation in these characteristics obeys a similar bilateral exponential (Laplace) law as well (see Fig. 1.4). Similar bilateral exponential laws describe the differences $\Delta V^2(t) = V^2{}_{max}(t + \Delta t_1) - V^2{}_{max}(t)$ and $\Delta P(t) = \delta P(t + \Delta t_2) - \delta P(t)$ obtained with time intervals of $\Delta t_1 = 12$ and $\Delta t_2 = 6$ h for catalogs 1 and 2, respectively. An interpretation of these laws will be given below where we represent the evolution of a typhoon by a cumulative (additive) model.

The next important example of natural disasters whose distribution is close to the exponential law is provided by the data on maximum annual river flow discharge. Figure 1.5 shows distributions of annual maximum discharge for several rivers. These data were obtained from the Institute of Water Problems of Russian Academy of Sciences, Moscow, courtesy to Dr. M. Bolgov. It can be seen that in most cases the data distribution are close to the exponential law. Significant deviations occur, however, for a few rivers where extremely large floods have been recorded. The distribution of these rare, extremely large events approaches a power-law.

These features agree with the character of rating suggested for the cases of moderate and major floods. The Myer rating for moderate floods has the form $K = Q/A^{1/2}$ where Q is the maximum discharge in m^3/s, and A is the basin area in km^2 [Wo]. But the log scaling K is used in rating the maximum observed floods [RR3, Wo]

$$K = 10[1 - (log(Q) - 6)/(log(A) - 8)] \tag{1.5}$$

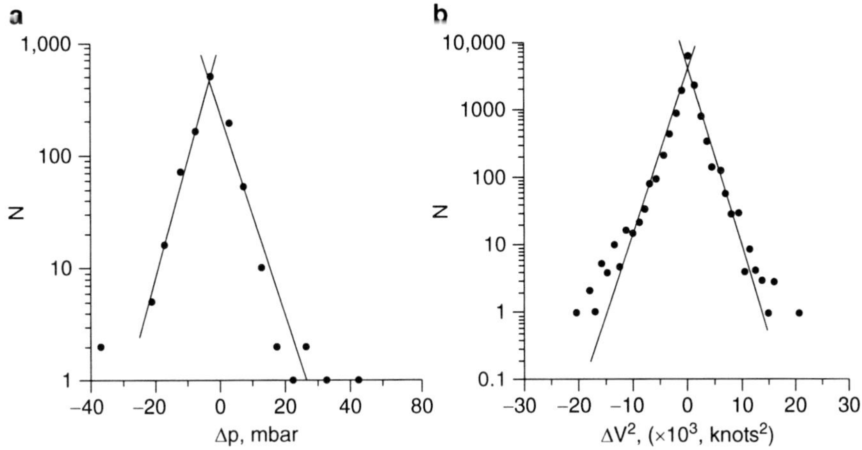

Fig. 1.4 Histograms of changes $\delta P(t+\Delta t_2) - \delta P(t)$ of pressure deficit δP in the center of hurricanes (**a**) and changes $V^2_{max}(t+\Delta t_1) - V^2_{max}(t)$ of squared maximum wind velocity (**b**) measured at regular time intervals Δt_2 and Δt_l, respectively; lines show fits by the bilateral exponential Laplace distribution

Fig. 1.5 Distributions of number N of annual maximum water discharge not less than J for the rivers Arkansas, 1895–1960 (**1**), Chirchik, 1900–1963 (**2**), Neman, 1812–1959 (**3**), and Danube, 1893–1960 (**4**). Fitted exponential distributions are given by *dotted straight lines*

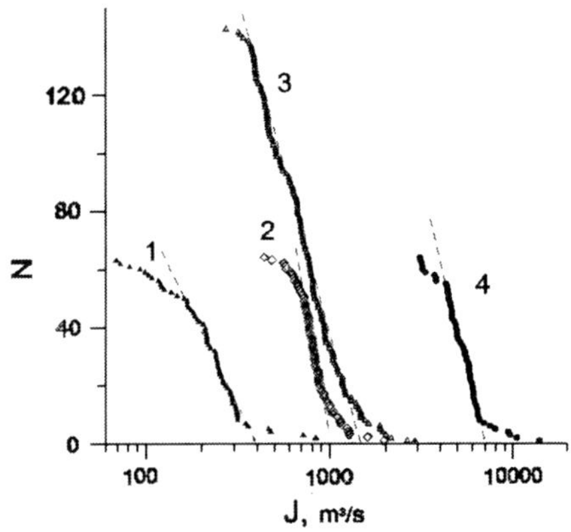

The linear and logarithmic scales used in the rating of moderate and largest floods provide an indirect support the appropriateness of the exponential and power-law distributions found above for such events.

One important feature of the exponential law that distinguishes it from the normal law is a much higher probability of occurrence for larger events. Still more extreme values are characteristic of another classical type of distribution to be discussed below, viz., the Pareto power law.

The power-law distribution is often encountered in very diverse physical systems. In this case the probability density $f(x)$ is

$$f(x) = b/x^{1+b}, \quad b > 0; \quad x \geq 1. \tag{1.6}$$

In different physical systems the size X refers to the energy or to the spatial "size" of the events or objects in hand. The power-law distribution is a "self-similar" distribution because of the following property it possesses. In a large population that obeys the Pareto law, the ratio of the numbers of events with approximate sizes A and aA, $n(A)/n(aA)$, does not depend on A. Here a is an arbitrary positive number. Thus, the power-law distribution is scale invariant (this is an expression of the principle of self-similarity).

Comparing the Pareto distribution with the normal and exponential distributions, one concludes that the decrease in the probability of large events is the slowest for the Pareto distribution. The contribution of the rare largest events into the total effect rapidly increases in a consequence of typical distribution laws from the normal law to the exponential law and to the power-law distribution. Moreover, for a power-law distribution with $b < 1$ the sum of a sample $(x_1 + x_2 + \cdots + x_n)$ is comparable with the maximum $x_{max} < max(x_1, x_2, \ldots, x_n)$. Distributions with this property cannot be characterized by their sample means. The theoretical mean (mathematical expectation) will be infinite. Even though based on a finite sample, the sample mean is unstable and tends to increase with increasing sample size. A real example showing that the mean does not approach a stable value with increasing sample size is given below in Table 1.1.

In recent decades self-similar power-law distributions were found for very different natural phenomena [Ma, TM, Tu, Wo]. The power-law distributions were shown to be typical of different types of catastrophic processes [She]. This fact has been definitely established for the case of seismicity, both for the physical parameters of earthquakes (seismic energy and scalar seismic moment [Kas, DES, So1]) and for the casualties and loss of property from earthquakes [Kas, RP1, RP3, PS1]. The seismic moment distributions for different types of earthquakes occurring in subduction zones with "strong" and "weak" interaction between the slab and the overlying mantle block, for deep earthquakes, and for the mid-oceanic rift zone earthquakes are presented in Fig. 1.6; all these empirical distributions are fitted fairly well by power laws.

The next important example is concerned with volcanic eruptions. The data source was the catalog [Gu2]. This catalog mostly contains eruption coordinates and times, but for a number of eruptions it also includes the volume of discharged material (total volume, lava, volcanic ashes) and/or energy estimates. The distributions of volumes of discharged material and released energy are well fitted by the power law (Fig. 1.7). We derived the regression of eruption energy on different kinds of discharged material. The additional estimates of energy release thus obtained were added to the catalog, resulting in a total of 286 energy releases for volcanic eruptions. This distribution and original volumes of discharged material and energy can be described by the Pareto law with $b = -0.7 \pm 0.3$. Very similar results both for the validity of the power-law distribution and the b-value were obtained in [SS2, VMGAO]. The statistical results presented here are supported by the logarithmic scale of tephra volume used in evaluation of Volcanic Explosivity Index VEI which is used to scale the size of volcanic eruptions [Wo].

Table 1.1 Mean annual estimates of losses for different averaging time intervals

	Earthquakes		Floods	
Length of intervals used for estimation of average loss, years	Significant earthquakes World data base[a] 1900–1996, b = 0.79 ± 0.02, thousands of casualties	Economic losses in the USA 1925–64, million \$, b = 0.85 ± 0.24 (Data from [DK])	Economic losses in the USA, 1925–64, million \$, b = 1.33 ± 0.42 (Data from [DK])	Number of homeless World data, 1964–1991, b = 0.76 ± 0.14, thousand people (Data from [Wa])
1	30 ± 96			From 50 to 28,000
5	31 ± 43			1,019
				1,213
				1,443
				2,067
				6,418
7	32 ± 80			932
				951
				2,245
				4,761
10	32 ± 38	2.54	267.8	
		13.06	322.0	
		13.8	410.4	
		43.62	451.2	
15	32 ± 26		337.8	1,683
			341.6	3,039
			377.5	
20	35 ± 27	5.22	319.2	
		10.05	333.3	
		39.08		

[a]This data base is described in detail in Chapter 4

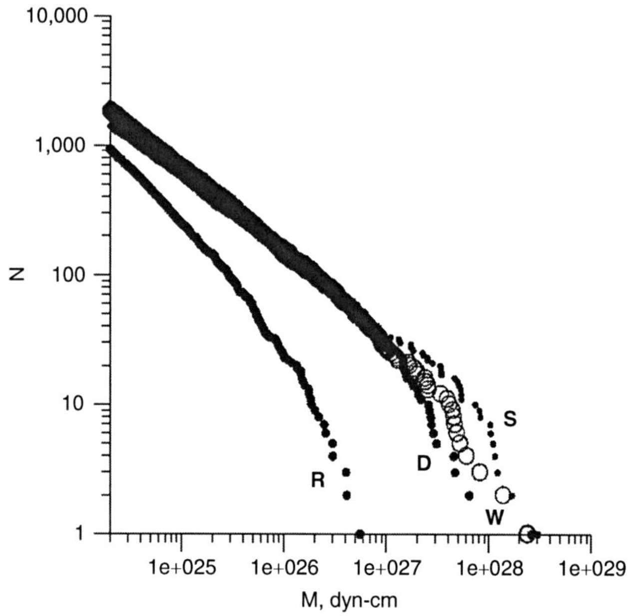

Fig. 1.6 A classical example of the power-law distribution of the number of earthquakes *N* with seismic moment value not less than *M*; R stands for earthquakes of mid-oceanic ridge zones, D for deep earthquakes (H > 70 km), S and W for shallow (H < 70 km) parts of subduction zones with strong (S) and weak (W) mechanical interaction with the continental block (classification of the subduction zones from [Ja])

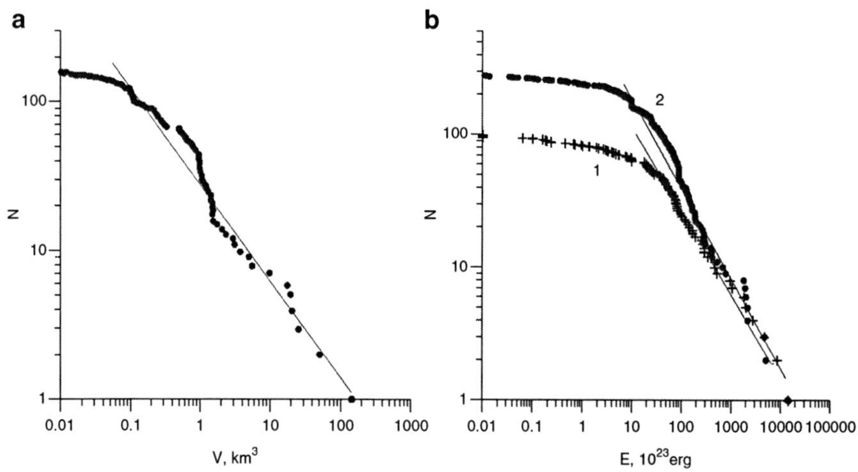

Fig. 1.7 Distributions of number *N* of volumes discharged by eruptions not less than *V* (**a**) and eruption energy values no less than *E* (**b**) from catalog [Gu2] (**1**) and with added values of explosion energy (**2**), see text for details. *Lines* give fitted power-law distributions

The next important example of a power-law distribution for natural disasters is connected with rare, extremely violent floods. Figure 1.5 presents the distribution of maximum annual river discharge for several rivers. In the moderate range the distributions are close to an exponential law. However, the distribution of rare, largest floods is different. The data for the Arkansas and Danube rivers are presented in Fig. 1.8 as a log–log plot. A break in the distribution law can be seen: the distribution of the greatest discharges is close to a straight line, showing that a power-law distribution is suitable for this range. The conclusion that the largest floods obey a power-law distribution agrees with results of geomorphologic analysis [An]. As shown in this work, the relief can contribute to the occurrence of extremely violent water streams. The occurrence of such streams is connected with changes in the systems of water reservoirs (failures in natural dams and escape of water from reservoirs). We remind the reader that, as has been mentioned above (1.5), the logarithmic scale is used in the rating of the largest floods.

Power-law distributions are also typical of many other types of natural disasters. According to [Sol, M2, SGK], the distribution of tsunami heights H obeys a self-similar power law. Though more detailed information for 1992–1998 leads to the conclusion about the best fit provided by the lognormal law, the power-law distribution was found to be more appropriate for several cases [PR]. The power-law character of the distribution of tsunami heights H made a log scale eminently

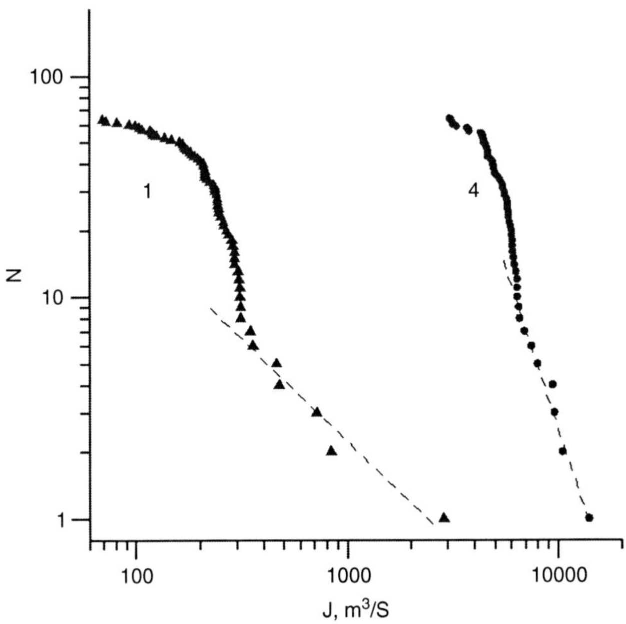

Fig. 1.8 Distributions of number N of annual maximum water discharges not less than J for the rivers Arkansas, 1895–1960 (**1**) and Danube, 1893–1960 (**4**). Fitted power-law distributions are shown by *dotted straight lines*

suitable for tsunami rating. The tsunami magnitude indexes I_s are defined [SGK, Ab2] as

$$I_a = a\,log(H) + b\,log(R) + c, \tag{1.7}$$

where R is the distance from the earthquake epicenter, and a, b. and c are coefficients differing in different scales.

According to [Yo, Wo], the power-law distribution also describes landslide size and the sizes of asteroid craters [Bar, She, Wo], forest fires [MM], and some other natural disasters.

Concluding this discussion of distributions typical of physical parameters of different natural disasters, we would like to stress two points. From the aforesaid one can notice a certain tendency toward changes in the character of distribution with increasing size of disaster. While the distribution of moderate-sized disasters can be described by the exponential (or lognormal) distribution, the distribution of rare largest events is closer to the power-law distribution. A model of such mixture of an exponential and a power-law distribution will be discussed later.

Note also that the variables (parameters) can be transformed to pass from one type of distribution to another. Thus, for example, if we use, instead of earthquake energy E (or the scalar seismic moment M), the magnitude m, then the power law will be replaced by the exponential law.

1.3 Distributions of Death Tolls and Losses Due to Disasters

Self-similar power-law distributions are typical not only of the physical characteristics of major natural disasters; this type of distribution is even more typical of death tolls and losses due to disasters. It was shown [HK, She, RS1, RS2, KS3, KPR1, Ry, RP1] that the empirical distributions of losses due to different disasters can generally be well fitted by the self-similar power-law distribution. This fit is better for large (causing heavy losses) events than for moderate-sized events.

The most complete death toll data for recent and historical natural disasters are available for the case of earthquakes, and the better quality data on economic losses are available for disasters that have occurred in the USA in the second half of the twentieth century (because of the well-developed insurance market).

To compile a data set of earthquake death tolls we used, in addition to the World Map of Significant earthquakes [M1] (http://earthquake.usgs.gov/earthquakes/eqarchives/significant, accessed in 2005) data bases, also materials from archives of large earthquakes of the world collected and processed by Nelson and Ganze [NG], [She], V.I. Khalturin (personal communication 2003). The aggregate of these data provides an extensive data base, and for a longer time interval, than is available for any other type of natural disasters.

Because of drastic changes in population density, the numbers of casualties are presented in two forms: the actual historical numbers of casualties and modified values as normalized by the recent population density. In both cases the death toll numbers could be well fitted by a power-law distribution (Fig. 1.9). We note, however, that the above modification is not fairly valid because of great changes in vulnerability over time (see Chapter 7 for details). That is why the actual death toll values are used below, unless otherwise specified. The statistics of earthquake death toll will be discussed in more detail below in Chapters 4 and 7.

A power-law distribution was found to fit the losses due to earthquakes, floods and hurricanes occurring in the USA during the last decades www://coastal.er.usgs.gov/hurricane_forecast/barton4.html, accessed in 2005). These data (Fig. 1.10) can be considered as the most reliable because of the development of the insurance market in the USA.

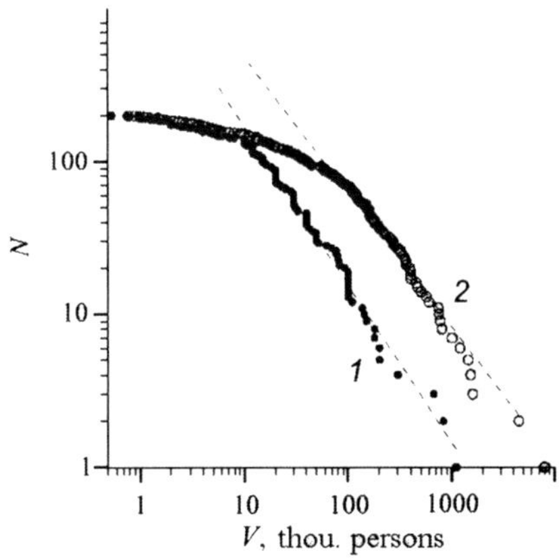

Fig. 1.9 Distributions of number N of major historical earthquakes with number of victims not less than V. (**1**) real number of victims, (**2**) number of victims normalized by the recent population. The fitted power laws are given by *dotted lines*

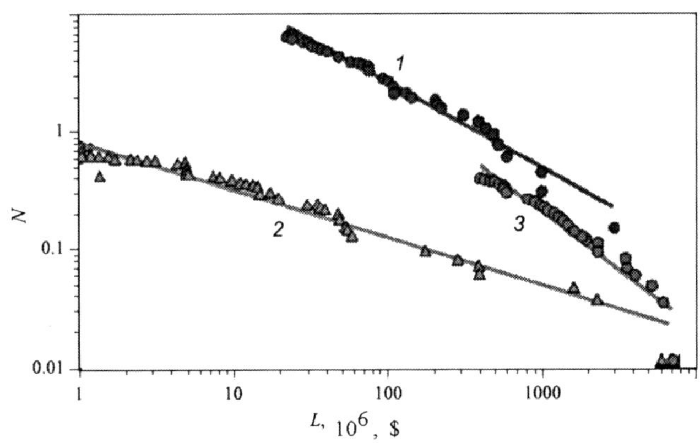

Fig. 1.10 Mean number N of events per year with economic losses greater than L: floods (**1**), earthquakes (**2**) and hurricanes (**3**); data for the USA for 1900–1989 (earthquakes and hurricanes) and for 1986–1992 (floods). The fitted power-law distributions are given by *lines*: $b = 0.74$ (for floods), $b = 0.41$ (for earthquakes), and $b = 0.98$ (for hurricanes). The figure was borrowed from www://coastal.er.usgs.gov/hurricane_forecast/barton4.html, accessed in 2005

From Fig. 1.10 two important conclusions can be drawn. Firstly, in all these three cases (earthquakes, floods, and hurricanes) the empirical distributions deviate from the pure power-law relation in the range both of largest and of smallest disasters. A possible cause of the deviation for the case of smaller disasters was discussed above, in Section 1.1. Deviations for the rare largest disasters will be discussed in Chapters 4, 5, 6, and 7.

Secondly, the power-law exponent (the b-value) shows noticeable differences between different kinds of disasters: $b = 0.98$ for hurricanes, $b = 0.74$ for floods, and $b = 0.41$ for earthquakes. Thus, the relative contribution from a few largest events in the total loss is the greatest for earthquakes and the lowest for hurricanes. This change correlates with the magnitude distribution for these kinds of disaster discussed above: the energy of earthquakes obeys a power-law distribution, for floods the power-law character of distribution is valid for the largest events only, whereas for moderate events the distribution approaches the exponential law; finally, the energy of hurricanes obeys the exponential distribution. This means that the contribution from the rare extreme events in the sum total energy of disasters is the greatest in the case of earthquakes and the lowest for hurricanes.

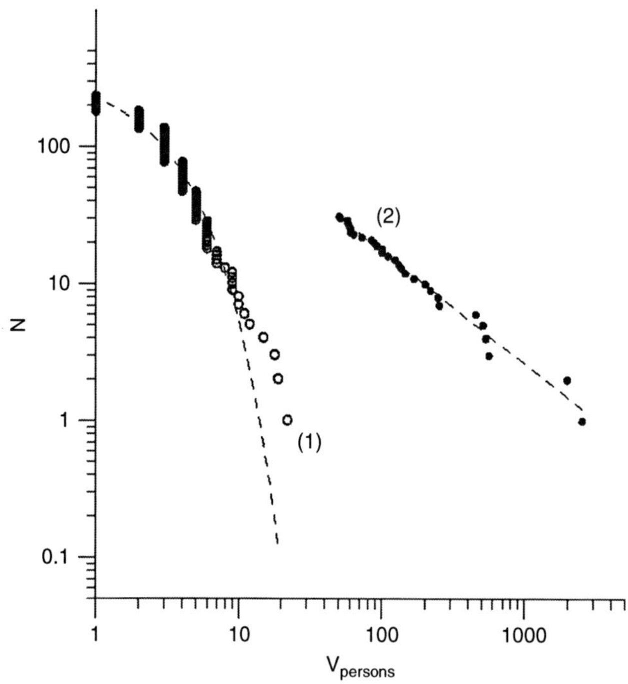

Fig. 1.11 Distribution of numbers N of accidents with death tolls not less than V from explosions and fires in the Russian Federation for 1992–1996 according to [SAK] (**1**), and from the largest manmade accidents in the world in 1900–1988 [Po], (**2**). The *dotted lines* give fitted distributions: exponential for (1) and power-law for (2)

The contribution from the largest events into the total economic loss are in the same order.

Concluding this brief review of loss data for natural disasters, it seems suitable to touch upon the case of man-induced disasters. Special interest attaches to the death tolls caused by major man-induced accidents occurring in the world in 1900–1988 [Po]. These data can be well fitted by a power-law distribution (Fig. 1.11). Data on the accidents that have occurred in the Russian Federation can be found in [SAK]. The distribution of the numbers of casualties due to moderate-sized accidents appears to be closer to the exponential distribution (Fig. 1.11). Thus, the distributions of death tolls due to moderate and major man-induced accidents are different, similarly to the case of natural disasters. In both cases the distribution of the largest events approaches the power-law distribution, whereas that of moderate-size events approaches the exponential law.

The results presented in Sections 1.2 and 1.3 show that it is possible to describe empirical distributions of physical characteristics of natural disasters and losses from disasters using three typical physical distribution laws: the normal Gaussian, the exponential Boltzmann, the Pareto power-law distribution, and combinations of these. This unified approach is alternative to the use of specific approaches (the traditionally used parametric distribution functions) for every kind of disasters. Application of these three distributions widely used in physics is justified by their transparent physical meaning, which enables the use of well-known physical analogues in interpreting the generation of natural disasters.

1.4 The Classification and Parameterization of Disasters

The variety of properties in different kinds of natural disasters and adverse phenomena is very wide. The situation in this field of natural sciences is described in "The mathematics of Natural Catastrophes" [Wo] as follows: "the sparsity and inhomogeneity of data, the range of extremes, and common divergence in the opinions of the Earth scientists". But for practical needs, a unified system of classification and parameterization of such phenomena is very desirable. To meet this demand a number of classifications of natural disasters and adverse phenomena have been proposed. It was proposed to classify the events into natural, man-made, and mixed, into fast and slow, by loss or energy, by the state of the operating agent (air, water, or rocks), and by the character of action (mechanical, thermal, chemical, or others). All these classifications have not however been widely used in practice, probably because of an inadequate choice of classification parameters which do not provide an effective classification for the wide range of natural disasters and adverse phenomena. Actually, all these classification schemes are descriptive in character. At the same time, the experience in the development of successful and widely used classifications shows that these are based not on the most prominent attributes of the phenomena to be classified, but on systemic attributes.

The important distinctions between the classical physical distribution laws were used in the scheme of classification and parameterization of disasters and adverse phenomena suggested by Kuznetsov et al. [KPR2, 03, R2]. The contribution of the largest events sharply increases as one goes from one distribution to another, in the sequence: the normal law – the exponential law – the power law. The level of instability of the processes that generate natural disasters increases in the same order. These properties were used as the basis for the classification scheme put forward here.

As the first main parameter for classification, the dimensionless ratio R of disaster size (magnitude in the usual terminology, or a parameter that represents it) to the average background value of the same parameter is proposed: $R =$ (typical disaster magnitude)/(background magnitude). The examples are: the ratio of water discharge during a flood to the average annual discharge; the ratio of peak amplitude in seismic waves of an earthquake to the amplitude of the microseisms.

All disasters are classified into three types depending on their R value. The sequence of increasing R values defines the following types:

- Disasters of the "trend" type
- Disasters of the "extremum" type
- Disasters of the "breakdown" type

The basic characteristics of these types are summarized in Table 1.2. It is easy to see that the type of a given kind correlates closely with the type of distribution law suitable for the description of magnitudes of this disaster:

- The Gaussian law
- The exponential law
- The power-law distribution

Table 1.2 The scheme of disaster classification

Characteristics of disasters	Disaster types		
	Trend	Extremum	Breakdown
R parameter	~1	1.5–5	$10 - 10^5$ and more
Typical examples	Climate and slow sea level change	Short-time sea level change, floods, droughts	Earthquakes, volcanic eruptions, major floods
Typical distribution law	Normal	Exponential	Power-law
Type of generating model[a]		Cumulative process	Multiplicative process
Parameters of statistical description[b]	Mean and variance, regression	Mean and variance	Order statistics, medians, quantiles
Most effective ways of loss mitigation	Preventive measures	Mixed approach	Rescue services and insurance

[a]This characteristic is discussed in Chapter 2
[b]This characteristic is discussed in Chapter 4

Examples of the trend-type disasters include climate change and long-term sea level variations. The extremum type disasters are moderate-term sea level variations, the droughts, and river floods. The breakdown type disasters are earthquakes, eruptions of volcanoes, meteorite impact, etc. For the disasters of the first type the R-value is somewhat more than unity, for the second type the R-values vary from 1.5 to 5–10, and for the third type the R-values vary from 10 up to 10^5 or even greater.

An exhaustive classification should characterize all kinds of disasters by a few parameters. We suggest characterizing each kind of disasters by two parameters: the intensive parameter R and the extensive parameter L (loss), which describes the characteristic loss from this disaster in a region. The concepts "extensive parameter" and "intensive parameter" are used here in the sense accepted in systems analysis [Gi]. Note that the parameterization of phenomena by the intensive and extensive parameters is considered in systems analysis as a very compact but informative form of description.

As a numerical estimator of the extensive parameter it seems reasonable to use the typical loss value L, or typical value of $logL$, for a given interval of time (10 years, say). The adjective "typical" does not necessarily mean here and below an "average" (arithmetic mean). Sometimes, for the extremum type disasters, and in particular, for the breakdown type disasters, the typical value of a sample should be estimated by the median of this sample, since this statistic is more stable. The passage to logarithms $logL$ can help as well, in order to avoid instability of estimates. The routine use of the sample mean can be incorrect for a considerable number of disaster types, specifically, in all cases where the data obey a power-law distribution with $b \leq 1$ (for examples see Table 1.1). The correct technique for estimating L in such cases is given in Chapter 3.

Despite the evident simplicity, this scheme for disaster parameterization by the (R, L) pair appears to be useful for summarizing information about different kinds of disasters. For example, the disasters of different types differ in the relative efficiency of the methods in use for loss mitigation. The parameters of the first disaster type evolve slowly, and the impact of these disasters can be rather well predicted. Hence it is natural that in this case preventive measures can be very effective for loss reduction. Dam construction against sea floods is an example showing the efficiency of preventive measures in the case of the first type disasters.

On the contrary, for the third type disasters the impact surpasses the background level by several orders of magnitude. The disaster evolves (at least, in its final part) very quickly. Its prediction is highly problematic, if not altogether impossible. Hence it is natural to assume that direct preventive measures will be of limited efficiency, especially in the case of major disasters. These considerations agree with the insurance practice; according to statistics by Dacy and Kunreuther [DK], preventive measures used in the USA provided rather poor (5–10%) loss reduction in the case of major hurricanes, whereas in the case of smaller hurricanes the reduction was as high as 60–80%. For the case of breakdown type disasters the most effective tools are the insurance and rescue services in *force majeur* situations.

Empirical distributions of disaster magnitude for a number of different kinds of natural disasters are given in Fig. 1.12 where they are presented on a log–log or on a semi-logarithmic scale. The distributions are seen to be essentially different. The empirical distributions close to a power law in the log–log plot can be well fitted by a straight line, whereas the empirical distributions that are close to a straight line on a semi-logarithmic scale can be well fitted by an exponential law. The normal distribution corresponds to a nearly vertical line in a log–log plot and to a parabola in semi-logarithmic coordinates. Thus, the character of the distribution, and accordingly, the type of disaster, and even some characteristic properties of a given kind of disasters, can be predicted with a certain reliability by the character of their empirical distribution curves.

From empirical distributions of disaster size the tendency mentioned above (a change in the distribution taking place with increasing magnitude) can be seen. This tendency is fairly evident for cases of floods (Fig. 1.7) and for the death tolls of man-induced disasters (Fig. 1.11). The situation appears to be less clear in some other cases (several curves in Figs. 1.12, 1.1, and 1.2). The tendency is as follows. While the distribution for smaller disasters is close to the normal law, and that of moderate-sized disasters is close to the exponential law, the distribution of the rare largest events tends to agree with the power law.

This tendency appears to have certain analogues (probably even an explanation) in theoretical statistics. As a matter of fact, the conditions for the normal distribution are probably satisfied in the case of smaller disasters, because all effects are then expected to be moderate in value. The occurrence of strongest disasters implies

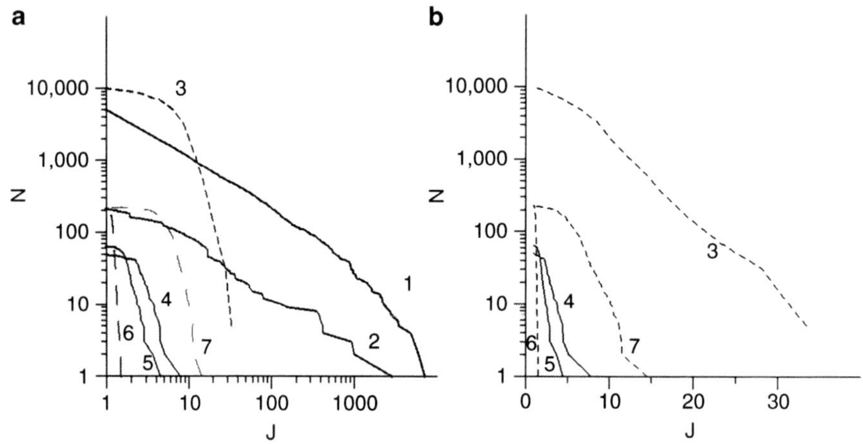

Fig. 1.12 Distributions in log–log (**a**) and in a semi-log (**b**) scale of numbers of events N with the parameter R not less than J. (**1**) Distribution of seismic moment (normalized by $M_{min} = 5 \times 10^{17}$ J); (**2**) energy of volcanic eruptions (normalized by $E_{min} = 5 \times 10^{19}$ J); (**3**) distribution of wind velocities at the South Kurilsk weather station ($V_{min} = 1$ m/s); (**4**) normalized maximum annual water discharges by the Araks river (1912–1959); (**5**) the same for the Chirchik river (1900–1963); deviations from mean temperature for summer (**6**) and winter (**7**) for St.-Petersburg, 1744–1989. Data sources are given in the text

the possibility of very high contributions; in these cases the other limit theorems of the theory of probability are suitable. But the limit distributions in these cases would be different from the normal law and will obey the power law distribution for the range of large values. The occurrence of a power-law distribution in the case of largest disasters can probably be explained in terms of the multiplicative models to be discussed in Chapter 2.

For practice it seems important that each change in the character of distribution should correspond to an increase in the probability of occurrence for the larger events. In practice, however, the probability of rare great disasters is often extrapolated from the distribution of smaller events. Such extrapolation can result in a substantial underestimation of the probability of large events (either man-induced accidents or natural disasters). The fact that such underestimation is possible is demonstrated by empirical data. It was noted in a review [VVM] that the actually observed frequency of major accidents often exceeds the expected probability by two and more orders. A simple multiplicative model taking into account synergetic effects and providing an increased probability of large events, is proposed in the following chapter.

To conclude this discussion of our disaster classification scheme, we note that the popular notion of a disaster as a sudden and outstanding event conforms to the third (breakdown) type disasters with large values of the parameter R. The large values of R are typical of disasters with power-law distributions. This type of distribution is typical of earthquakes, volcanic eruptions, and the larger floods. The power-law character of distribution is also typical of the death toll and loss values due to different kinds of natural disasters. From the aforesaid the special importance of heavy-tailed power-law distributions is evident. Most of this monograph is devoted to the study of heavy-tailed power-law distributions and statistical approaches applicable to such distributions.

1.5 The Main Results

The empirical distributions of the physical parameters that characterize natural and man-induced disasters and the distributions of the losses caused by these disasters are rather different. Nevertheless, all these distributions can be adequately described by three classical physical distribution laws: the normal Gaussian law, the Boltzmann exponential law, and the Pareto power-law distribution, or by their combinations. This approach is an alternative to the use of disaster-specific parametric distributions. The application of these three classical physical distributions is justified by their transparent physical meaning which permits the use of physical analogues and models in interpretation of disaster generation.

The qualitative distinction between the normal, exponential, and power-law distributions reflects a basic distinction in the characteristics of different kinds of natural disasters. These basic distinctions underlie our three-type classification scheme of natural disasters. These types are the following: the disasters of trend, the disasters of extremum, and the disasters of breakdown. The dimensionless parameter R, which is the ratio of disaster magnitude to the background level is used to classify disasters as belonging to one of these three types. Naturally, the R-value and the distribution type of its magnitude are closely connected. The classification here proposed appears to be useful in the planning of loss mitigation measures.

The greatest threat is posed by the disasters of the third type, disasters of breakdown type. Such disasters cause the greatest damage. The distribution law typical of the size of these disasters and of the losses due to disasters is a power-law distribution. In a number of cases these laws have an infinite mean (mathematical expectation), and the routine statistical approach based on sample means and variances is inappropriate for such distributions. The statistical approaches suitable for such distributions are discussed in detail in the chapters to follow.

Chapter 2
Models for the Generation of Distributions of Different Types

One of the principal objects of theoretical research in any department of knowledge is to find the point of view from which the subject appears in its greatest simplicity

J. Willard Gibbs

Everything should be made as simple as possible, but not simpler

A. Einstein

In the previous chapter it was shown that three distributions which are widely used in physics, viz., the normal Gaussian distribution, the exponential (Boltzmann) distribution, the Pareto power-law distribution, and combinations of these, are sufficient to describe many empirical distributions of physical characteristics of natural disasters and of losses due to disasters. The use of these typical distributions for fitting the empirical distributions facilitates data interpretation and promotes the best understanding of the processes underlying the generation of disasters. Similarity of distributions in physics and in the study of natural disasters is of help in understanding the latter. In this chapter we discuss general physical and mathematical models capable of explaining the processes of generation of the abovementioned typical distribution laws.

2.1 Why Are the Characteristic Types of Distribution Prevalent?

The characteristic physical distributions listed above are very common in nature. What are the properties that cause so frequent an occurrence of these distributions? The case of the normal law seems to be the most simple. The prevalence *of the normal distribution* can be treated as a consequence of the Central Limit Theorem (CLT) in probability theory. The normal distribution is found in all cases where the result can be treated as a sum of contributions from a large number of independent factors with finite variance, none of the factors being much larger than any other. If the involved factors are interrelated, or there is a predominant factor, the

V. Pisarenko and M. Rodkin, *Heavy-Tailed Distributions in Disaster Analysis*, Advances in Natural and Technological Hazards Research 30,
DOI 10.1007/978-90-481-9171-0_2,

distribution of the sum can differ from the normal law, first of all, in the range of extreme values. Note also that the normal law meets the condition of maximum uncertainty (maximum entropy) in the case when the values of two first statistical moments (i.e., mean and variance) are given.

The prevalence of the conditions for generation of the normal distribution often gives rise to the illusion of universal applicability of the normal law. In this connection H. Poincaré noticed that the physicists trust in the normal law because they assume it to be a mathematically proven conclusion, whereas the mathematicians trust in the normal law because they think it is a physical law. This remark of Poincaré's remains as apposite as ever. An obvious or implicit assumption of applicability of the normal law exists really in all cases where the sample mean and variance are used to describe observational data. However, the use of these characteristics in cases where the empirical data are widely divergent from the normal law leads to incorrect inferences. An example of such aberrant application of the normal law was presented above (Table 1.1) where the sample means are subject to large scatter and do not approach any stable value, no matter how long the time interval of averaging is used.

The Boltzmann distribution represents another typical example of a physical distribution. The probability density of this law has an exponential form:

$$f(E_i) \sim exp(-E_i/\theta), \tag{2.1}$$

where, according to the usual interpretation, the E_i are the energies of components (particles) of the system in thermal equilibrium with a thermostat of temperature θ (commonly denoted $\theta = kT$, where T is the temperature and k the Boltzmann constant).

The exponential distribution meets the condition of maximum uncertainty (maximum entropy) when only the first moment (i.e., the mean equal to θ) is fixed. In the usual physical interpretation this law describes a random distribution of sample energy values in a thermal equilibrium. For example, it describes the distribution of energies of gas molecules in equilibrium with a thermostat of a fixed temperature.

In the case of natural disasters the energy of a disaster cannot generally be evaluated directly. For this reason the distributions of other, more easily determinable parameters are considered. When the distribution of a given parameter is found to be close to the exponential law, it seems reasonable to suppose that this parameter is linearly related (in the statistical sense) to the disaster energy. In that case the corresponding θ value can be evaluated, and this value will characterize the "temperature" of the system that generates the ensemble of disasters under consideration.

This approach was found to be useful for the treatment of disasters. For example, it was shown for hurricanes that the pressure deficit at the center of a hurricane and the squared maximum wind velocity are strongly correlated, and both the distribution of deficit of pressure and the squared maximum velocity can be well described by the exponential law, see Fig. 1.3 [GPRY]. From the physical point of view, both of these characteristics are closely connected to the energy of the hurricane. The hurricanes can be treated as "particles" that are in thermodynamic equilibrium with a certain reservoir. Moreover, both the changes of pressure deficit at the center of hurricanes and the squared maximum wind velocity obey the bilateral exponential (Laplace) distribution, see Fig. 1.4. The "temperature" θ of the "reservoir" was found to change slowly with time, and these changes appear to correlate with the intensity of the El-Nino phenomenon.

Quite similar results were found from an examination of the variability shown by economic indices. The short-time and long-time statistics of currencies and economic indices have been the subject of very extensive research (see, e.g. [Shi, MS] and references therein). Here, it is only one feature specific to these data sets which will be touched upon in connection with our examples of prevailing typical distribution laws. The daily 1-minute changes in the Dow Jones Industrial Average (DJIA) index were examined as an example (http://www.finam.ru/analysis/export/default.asp accessed in November 2009) and have been found to agree quite well with the Laplace distribution, at least around the center. Examples of this type of behavior for a few randomly sampled days are presented in Fig. 2.1. It can be seen that the distribution of differences of the 1-minute index values x_k (except for occasional extreme differences) is fairly well consistent with the Laplace law, and thus the "effective temperature" $\theta_t(T)$ can be defined in accordance with (2.1):

$$\theta_t(T) = (1/T) \sum_{k=t}^{T+t-1} |x_k - x_{k-1}|. \tag{2.2}$$

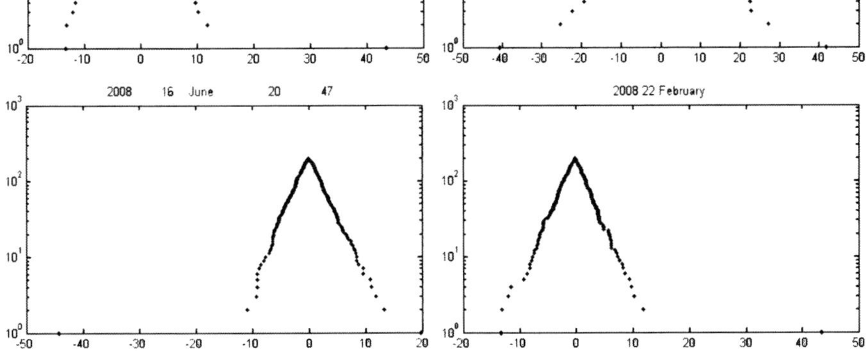

Fig. 2.1 Bilateral exponential (Laplace) distribution of differences ($x_k - x_{k-1}$) of 1-minute Dow Jones Industrial Average (DJIA) index (Data for four randomly sampled days are shown)

The time span T can be chosen according to the time scale of interest. We shall consider the daily effective temperature, where the interval $(t;\ t+T)$ covers a particular day, but other time spans T are of course possible (week, month, quarter, year). The notion of "effective temperature" is close in sense to the term "volatility" well known in economics. It is known that the volatility grows during periods of market instability.

Using Eq. 2.2, the daily "temperature" values for both positive and negative differences in the DJIA index were calculated in relation to t and compared with the current index values (Fig. 2.2a, b). It can be seen that the effective temperatures for positive and negative differences are very similar, and that episodes of substantial increase in the "temperature" tend to coincide with time intervals of a substantial change in the DJIA index.

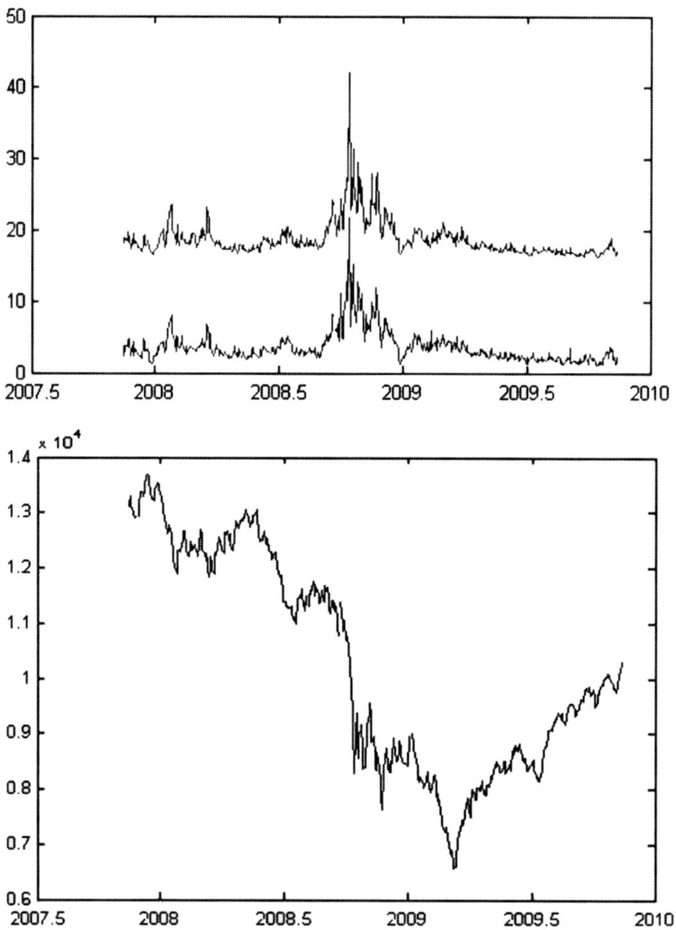

Fig. 2.2 "Effective daily temperature" for positive and negative differences of DJIA index displaced by 15 units along the y axis for convenience (**a**), and time series of the index value (**b**)

These examples of the bilateral exponential distribution for economic indices and hurricanes demonstrate the fruitfulness of the approach advocated here. The effective temperature thus defined appears to be a useful parameter for studying the behavior of dynamic systems not far from equilibrium.

The mechanism discussed above for the occurrence of the exponential law is not unique. The following simple models can give rise to an exponential distribution and can also help in interpretation of some kinds of disasters. Let us assume that the energy released by a disaster is equal to the energy that flows into the system at a constant rate during a preceding time interval. Suppose a disaster occurs when a certain trigger becomes active. If the times at which the trigger is activated are random (the Poisson process), then the time intervals between successive disasters will follow an exponential law. The energies released are thus proportional to the time intervals between disasters, and hence they will obey the exponential distribution. This model can be generalized to the case where the rate of energy inflow can fluctuate. Note that similar mechanisms of constant inflow of energy into a system with a subsequent sudden energy release are assumed in a number of models for disaster occurrence. Such models are widespread for example in modeling the time series of earthquakes.

Another model, which also gives rise to an exponential distribution, is the following. Suppose that the strength of a system consisting of a set of units is determined by the strength of its weakest unit. For simplicity we shall consider the one-dimensional case (a chain of links). We shall denote by $U(x)$ the probability that the strength of a chain of length x is sufficient to withstand some fixed load. The chain of length $(x + y)$ withstands the load if it is withstood by both the x and y sections. Believing that the strengths of the x and y sections are independent, we get the result that the probability $U(x + y)$ satisfies the equation

$$U(x + y) = U(x) \cdot U(y), \quad x, y > 0. \tag{2.3}$$

It follows from (2.3) that $U(x)$ obeys the exponential law

$$U(x) = exp(-kx),$$

where k is a constant. Hence, the probability density $f(x)$ for the lengths of the sections under a given load satisfies the exponential relation

$$f(x) = k \times exp(-kx), \quad x \geq 0. \tag{2.4}$$

From the above it can be seen that both mathematical and physical arguments can be advanced to explain the prevalence of the normal and exponential distribution laws in empirical disaster data.

Both mathematical and physical arguments can also be advanced to explain the prevalence of the power law distribution. The popular belief is that "in a statistical description of disasters the occurrence of the power law is a rule that has nearly no exceptions" [VVM]. Though it seems that this statement is too categorical, the wide prevalence of power-law distributions in the case of disasters is undoubted. This distribution law is especially typical of loss values. Moreover, a power-law distribution of losses takes place even for disasters whose physical parameters obey a distribution that is different from the power law (e.g., for hurricanes).

We begin with mathematical reasons for the power-law distribution. These reasons are based on the limit theorems in probability theory, in particular, on the theory of stable laws. First, we will show that a power-law distribution can be observed for major disasters. It follows from the Gnedenko–Pickands–Balkema–de Haan limit theorem (see Chapter 6 for more detail). This theorem asserts that the conditional distribution function $F_h(x)$

$$F_h(x) = P\{X - h < x|X > h\}$$

of excess values above some threshold h tends, as $h \to \infty$, to the Generalized Pareto Distribution (GPD) for any initial distribution function $F(x)$ that satisfies some general conditions. The GPD distribution function denoted as $G((x{-}h)/x,\ s)$ is written as follows:

$$G((x-h)/\xi, s) = 1 - (1 + \xi(x-h)/s)^{-1/\xi}; \quad x \geq h; \quad 1 + \xi(x-h)/s \geq 0, \quad (2.5)$$

where ξ, s are some parameters and h is a threshold.

According to this theorem, the excess distribution $F_h(x)$ has a power-like tail if $\xi > 0$. Having in mind this limitation, one can expect a power–law-like tail distribution for the greatest disasters.

The second general mathematical argument explaining the prevalence of power-law distributions is based upon the theory of Levy's stable laws (see [Z1, Z2]). We remind the definition of a stable distribution. Independent, identically distributed random variables $X_1, X_2, \ldots X_n$ obey a stable distribution law if the sum $(X_1 + \cdots + X_n)$, on being normalized by a constant C_n and centered by another constant A_n, has the same distribution law as the initial distribution of X_i. The normal Gaussian law is a special case of the stable laws. In this special case the constant C_n is proportional to $n^{1/2}$. Generally, we have $C_n = n^{(1/\alpha)}$, where $0 < \alpha \leq 2$. Thus, for the normal distribution one has $\alpha = 2$. The Cauchy and Levy distributions with $\alpha = 1$ and $\alpha = 1/2$, respectively, are other cases of Levy stable distributions. It is important that all Levy stable laws, except for the normal law, have a power-like tail. It is only the stable Levy distributions which can be the limit distributions of independent, identically distributed random variables.

But the effect due to a major disaster is the sum of a large number of local effects. If these effects have infinite variances (which is quite possible), then the CLT is not applicable, and the limit distribution of the total effect can only belong to the family of the stable laws with $0 < \alpha < 2$ and possesses a power-like tail. The special case of the normal distribution can be expected in the case of losses due to minor disasters whose local effects have finite variances.

These formal mathematical deductions agree fairly well with empirical distributions of disaster characteristics discussed above. In a number of cases we could see a power-like distribution for the rare largest events and the normal-like law for small events.

The power-law distribution has the form

$$P\{X > z\} \equiv 1 - F(Z) \sim C/z^a, \tag{2.6}$$

where C is a constant. It can be easily shown that power-law distributions have infinite variances for $\alpha < 2$ and infinite means for $\alpha \leq 1$. This fact causes serious difficulties for the calculation of cumulative sums of such effects. These difficulties will be discussed in detail in Chapter 4.

Following [ZS, Z1] we present a wide class of power-law distributions that emerge in many physical applications. Consider a set of point objects (particles, defects, etc.) distributed absolutely randomly in some volume (a Poisson point field). Each particle creates "a field of influence" described by a function of influence $U(X, Y)$, where $U(X, Y)$ describes the influence at point Y created by a particle located at point X. The total effect from different particles is equal to

$$U(Y) = \sum_{k=1}^{N} U(X_k, Y) \tag{2.7}$$

where N is the number of particles, defects of structure, etc. Such an "influence field" occurs as the gravitational field of point masses in space, as the stress field due to defects in a solid and in other physical systems. In a typical case, the function of influence $U(X, Y)$ is Green's function for the differential equation which describes the associated field. Green's function frequently has a power-law character (for the Laplace equation, for the heat conductivity equation, and in many other important cases). In all these cases $U(Y)$ obeys a stable distribution.

The Brownian motion model is another commonly occurring example of a power-law distribution. The time T_h of the first passage across some fixed threshold $h > 0$ along a Brownian trajectory issuing from zero obeys a Levy distribution with density

$$f_h(x) = h(2\pi x^3)^{-1/2} \cdot exp(-h^2/2x), \quad x > 0;$$

$$\text{i.e., } f_h(x) \sim x^{-3/2}, \quad x \to \infty. \tag{2.8}$$

When dealing with disasters, relation (2.8) can be used to describe a situation where a disaster occurs if the accumulation of stochastic damage effects reaches some specified threshold. It can be deduced from (2.8) that the mean T_h is infinite, though the event (the passage across a threshold h) occurs with probability one.

Below we examine two more models of stationary time series possessing a heavy tail.

Model 1. Suppose a time series is generated by the recurrence relation

$$Y_t = a_t Y_{t-1} + \varepsilon_t; \quad t = 1, 2, \ldots \tag{2.9}$$

where the ε_t are independent, identically distributed random variables (random innovations). This scheme is similar to the autoregression model of the first order (the Markov process), but unlike the Markov process where all the a_t are constant, $a_t \equiv a$ (the stability of a Markov process requires $/a/<1$), we shall assume that the a_t are random variables satisfying the following conditions:

- The $\{a_t, \varepsilon_t\}$ pairs are independent, identically distributed two-dimensional random variables
- $E log|a_t|<0$
- There exists such $\mu > 0$ that $0 < E\,/\varepsilon_t/^{1\mu} < \infty$
- $E|a_t|^{\mu} = 1$
- $E|a_t|^{\mu} log^{+}|a_t|^{\mu}<\infty$

Here the following notation is used:

$$log^{+}(x) = \begin{matrix} log(x), & if \;\; log(x) \geq 0; \\ 0, & if \;\; log(x)<0. \end{matrix}$$

Under these conditions, there is a stationary solution Y_t of (2.10) and its tail has an asymptotic power-law distribution (see [Ke]):

$$P\{|Y_t|>x\} \propto 1/x^{\mu}. \tag{2.10}$$

In the case $\mu < 1$ this distribution has a heavy tail. In this scheme a_t is less than unity "on the average" (more strictly, $E\ log\ /a_t/ < 0$), but nevertheless, because of possible occurrence of accidental series of large values $a_t, a_{t+1}, \ldots a_{t+K}$, there can be an "excursion" of the process $|Y_t|$ toward larger values. These excursions are observed, whenever extremely high values occur in the Y_t sequence. This results in a heavy tail. We remark that both in this case and in the examples above, a heavy tail appears because of a product of factors greater than unity. This cause of a heavy tail is apparently rather typical. With reference to the distribution of disaster casualties, this scheme corresponds to a synergetic effect from different adverse factors which are capable of substantially increasing the number of victims. For example, in the case of earthquakes such factors could include: the night time of disaster occurrence, bad weather conditions and other unfavorable phenomena that can accompany an earthquake: fires, dam failures, landslides, tsunami, etc. The combined synergetic effect of such factors can considerably increase the number of victims and economic losses.

The model (2.9) with $a = 1$ represents, as a matter of fact, the summation of effects ε_t:

$$Y_t = Y_0 + \varepsilon_1 + \cdots + \varepsilon_t. \tag{2.11}$$

It can thus be called **the additive (cumulative) model**. We are going to use it and compare it with the multiplicative model introduced below.

Model 2 is the ARCH process (Autoregressive Conditional Heteroskedasticity process) which was introduced into economics research by the Nobel Prize winner R.F. Engle [En].

In the elementary case the model is

$$Y_t = \left(\beta + \lambda Y_{t-1}^2\right)^{1/2} \varepsilon_t,$$

Or, which is the same:

$$Y_t^2 = \left(\beta + \lambda Y_{t-1}^2\right) \varepsilon_t^2, \tag{2.12}$$

where β, λ are parameters and e_t is white noise.

As was shown in [EKM, Theorem 8.4.12], if $0 < \lambda < 3.561$, then there exists a stationary solution of the recurrence relation (2.12) with a power-law tail:

$$P\{Y_t > z\} \sim 1/z^{2k}, \tag{2.13}$$

where k is the unique positive solution to

$$(2\lambda)^u G(u + 0.5) = \pi^{1/2},$$

where G is the standard gamma function. It appears that, when $\lambda > 1.57$, then the exponent in (2.13) is $2k < 1$, and so Y_t follows a distribution with a heavy tail. We see that the product of random variables $Y_{t-1}^2\ \varepsilon_t^2$ involved in (2.12) plays the same role as the product of random variables $a_t Y_{t-1}$ in (2.9): it can lead to "excursions" into the domain of large values which causes a heavy tail in the distribution. The above explains a wide occurrence of typical physical distributions discussed in Chapter 1.

We should also mention a cascade model that leads to one frequently occurring distribution, namely, the log-normal distribution. Models of this type were suggested for such natural phenomena as turbulence, the fragmentation of rock particles, tsunami waves (see [SBP, Sor, PR1]), and others. The energy transport in such processes is effected by a step-by-step transfer of the same mechanism from larger particles (whirls or other objects) to smaller particles (whirls). This is the so-called the "direct cascade" model (see [Ko, Kol, Mo, Fr]). The "inverse cascade" model was also suggested for seismicity (see [Kal, Shn]), where the process proceeds from small objects (cracks, geological faults) to larger objects. The total effect (energy of dissipation in turbulence, particle size in rock fragmentation) is determined by a product Π_n of positive random individual factors x_k: $\Pi_n = x_1, \ldots, x_n$. If we take logarithms, then we get:

$$log(\Pi_n) = \sum_{k=1}^{n} log(x_k).$$

Under some general conditions the limit distribution of this normalized sum tends to the normal distribution, so that the total effect Π_n has for its limit the log-normal distribution with probability density

$$f(x) = [2\pi\sigma^2 x^2]^{-1/2} exp(-(log x - m)^2/2\sigma^2); \quad x > 0,$$

where m, σ^2 are mean and variance of a normal distribution. The tail of a log-normal distribution is heavier than any exponential tail, but lighter than any power-law tail. All moments of the log-normal law are finite.

Finally, we consider the model where a heavy, power-law tail of the total effect is the result of multiplication of individual effects, viz., **the multiplicative model**.

Suppose that the total effect $\Pi_n = a_1 \times \cdots \times a_n$ is formed as a product of factors $a_k \equiv$ a, where the number of factors n is a random variable that obeys the geometric distribution

$$P\{n = k\} = (1 - p)p^k; \quad k = 0, 1, 2, \ldots, \quad 0 < p < 1.$$

The probability of a tail is here

$$\begin{aligned} P\{\Pi_n > x\} &= P\{a^n > x\} = \sum_{a^k > x} (1 - p)p^k = \sum_{k > \log x/\log a} (1 - p)p^k \\ &\cong p^{logx/loga} = (1/x)^{log(1/p)/log\, a}. \end{aligned} \tag{2.14}$$

We assume that $\log a > 0 \ (a > 1)$ and $log(1/p)/log\, a < 1$. Then the tail (2.14) will be "heavy", since the power exponent is $\{log(1/p)\ /log(a)\} < 1$. The heavy tail has appeared here due to a product of a random number of factors larger than unity. It is possible to generalize this scheme, considering stochastic independent, identically distributed a_k with a mean greater than $1/p$. In this case the asymptotic expression (2.14) holds true under some additional constraints on the distribution of a_k.

The random variable n which follows the geometric distribution may be looked upon as a "random process" starting at time $t = 0$ and ending at a random time $t = n$. The time can be a continuous random variable T, but in this case instead of the geometric distribution we have the exponential distribution with the distribution function

$$F(x) = P\{T < x\} = 1 - exp(-x/\tau), \quad x \geq 0, \tag{2.15}$$

where τ is a parameter. The corresponding time process can be considered as a Poisson process with intensity $1/\tau$ starting at time $t = 0$ and ending at the time of the first jump of the Poisson process. This process can occur as a sequence of independent Bernoulli trials with the probability of success (continuation) $p = 1 - dt/\tau$ and the probability of failure (stoppage) $1 - p = dt/\tau$. For continuous time we have

$$\Pi_T = exp(\alpha T),$$

where α is a positive parameter. We have

$$P\{\Pi_T > x\} = P\{exp(\alpha T) > x\} = P\{T > log(x)/\alpha\} = exp(-log(x)/\alpha\tau)$$
$$= 1/x^{1/\alpha\tau}. \tag{2.16}$$

If $\alpha\tau > 1$, then Π_T has a heavy tail. The analogy with the discrete case (2.14) is complete.

2.2 The Multiplicative Model of Disasters

Multiplicative models can be used for modeling disasters of different kinds. The parameters (p,a) in (2.14) or (τ,α) in (2.16) enable a power-law distribution to be modeled. This power distribution is characterized by a slope parameter β:

$$\beta = log(1/p)/log(a); \quad \text{discrete case;} \tag{2.17}$$

$$\beta = 1/\alpha\tau; \quad \text{continuous case.} \tag{2.18}$$

Parameter β characterizes the slope of the frequency-size plot of disasters on a log–log scale. As will be shown below, the parameter β is an analogue of the b-value in the Gutenberg–Richter frequency–magnitude relation.

Multiplicative models were shown to be useful for interpreting the nature of disasters of different kinds [R1, R3]. Because of the best availability of seismic data, we shall begin with a multiplicative model for earthquakes. As has been mentioned, the multiplicative model generates a power-law distribution similar to the Gutenberg–Richter relation (if earthquake size is described by the scalar seismic moment or seismic energy). Moreover, the multiplicative scheme has in this case a natural physical interpretation. According to one popular point of view [Gu3, Ra, RK2, MMS, etc.], the earthquake process can be treated as a sequence of episodes of failure of rigid inclusions (asperities). The system of rigid asperities, as well as geophysical media in general, has a hierarchical structure (see [Sa, SBR, Tu]). In accordance with this point of view, continuation or termination of the seismic process depends on whether the next rigid asperity of corresponding rank will be destroyed or not. A similar process is modeled by relation (2.13).

The multiplicative and cumulative models can be used for simulating various dynamic systems. It was shown in [GPRY] that some parameters related to the energy of tropical hurricanes obey the bilateral exponential (Laplace) distribution. However, no interpretation of this fact was suggested. In terms of the cumulative (additive) model this relation seems natural to be treated as a result of successive interactions between a hurricane and its environment (resulting in either growth or loss of hurricane energy). Such an interpretation would suggest new questions concerning interactions between hurricanes and their environment in various parts of the World Ocean.

A similar situation takes place with the application of the multiplicative scheme in econophysics. Financial data can be fitted by the bilateral exponential (Laplace) distribution (Fig. 2.1), hence the notion of "effective temperature" (2.2) can be defined in financial statistics relevant to some time interval T. We believe that this parameter can be useful when studying the behavior of world markets.

We would also like to note here a distinction between the distribution applicable to earthquakes and hurricanes. Both processes involve release of energy from active media: release of elastic energy of the lithosphere in the case of earthquakes and release of the thermal energy of heated subsurface sea layers in the case of hurricanes. One could suppose that the distributions of earthquake energy and hurricane energy are similar. However, this is not the case: the distribution of earthquake energy obeys the Gutenberg–Richter power-law relation, whereas the distribution of hurricane energy obeys the exponential Boltzmann distribution [GPRY]. Having in mind the multiplicative model, it is reasonable to hypothesize that this distinction is due to the avalanche-like evolution of failures in the case of earthquakes, which is absent in hurricanes. This hypothesis finds some confirmation. The analysis of empirical data on the dynamics of hurricanes has shown [GPRY] that the size of the central area of hurricane, where the pressure is the lowest, does not increase with decreasing pressure at the center of the hurricane. This feature contrasts sharply with the fast growth of the rupture zone with increasing earthquake magnitude. We also note that neighboring hurricanes do not tend to coalesce, in contrast to earthquakes where neighboring rupture zones generally coalesce. These features show that in the case of the hurricanes the condition $\dot{x} = kx$, $k>0$ is not fulfilled, and accordingly the power-law distribution does not occur.

An avalanche-like evolution is highly probable for forest fires and for snow and rock avalanches. Indeed, for these disasters power-law distributions are typical [She, Ba, TM].

2.3 The Mixed Models

We now discuss a somewhat more complex case concerned with empirical distributions of physical parameters of disasters and associated losses. We shall compare the death toll data from road accidents that had occurred in the Russian Federation from November 1993 until October 1994 [Saf], from fires and explosions that had occurred in the Russian Federation in 1992–1996 [SAK], and from the largest accidents in the world for the period 1907–1988 [Po]. These data are for different time spans. For the comparison to be valid, these empirical distributions were reduced to 5-year periods, i.e., all data were converted to the mean number of events occurring in 5-year time intervals. These values are presented in Fig. 2.3 as a log–log plot as numbers of events N with casualties not less than V persons.

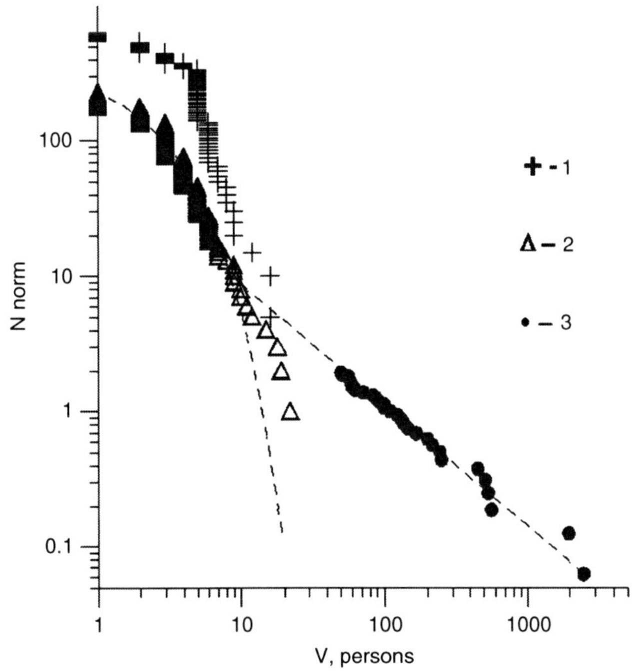

Fig. 2.3 Distribution of number of events N_{norm} normalized to the 5-year interval of time with numbers of victims not less V persons from manmade accidents. (**1**) Data on victims in road and transport incidents, from [Saf]; (**2**) data on victims due to fires and explosions in Russian Federation [SAK]; (**3**) the largest manmade accidents of the world [Po]. *Dashed lines* give the respective fitted distribution laws

For the smaller range the empirical distributions can be fitted by an exponential law shown by the dotted curve. However, for the greater range the data deviate from this exponential law, the number of such events essentially exceeds the fitted exponential function. It is also easy to see that the exponential distribution is inapplicable to the case of the greatest man-induced accidents of the world. When plotted on a log–log scale, this distribution is almost a straight line and accordingly, it can be fitted by a power law, which is shown in Fig. 2.3 as a dotted line.

Thus one can conclude that the distributions of death tolls due to moderate and to greatest man-induced disasters are different. The former can be fitted by an exponential law; accordingly, the mechanism operative in the evolution of such events can be modeled by an additive model. In contrast, the distribution of the greatest accidents agrees with a power-law distribution, and hence the evolution of such accidents should be modeled by a multiplicative model.

The same change in the character of distribution was found in the previous chapter for a few physical parameters of natural disasters (see Figs. 1.8 and 1.12). It was shown that the distribution of extremely large floods differs strongly from that of moderate size floods. The probability of occurrence of the greatest floods at several rivers is markedly higher compared with the expected from extrapolation of distribution law of moderate floods. Such an increase could occur as a result of a positive feedback between the size of a flood and the rate of its increase. This positive feedback can result from the breaking of various obstacles by the water flow (failure of dams, rectification of flow paths, etc.).

From the above it seems likely that the presence of a strong positive feedback between disaster factors will tend to make the distribution law more heavy-tailed. In such situations the empirical distribution throughout the whole range can be represented as a mixture of the exponential and the power law. The processes that give rise to these distributions were modeled above by the cumulative and the multiplicative scheme, respectively.

A simple and physically transparent mixed model can be described as follows. We assume that, with probability p_0, the size of the next disaster z_k follows the distribution function $F_1(x)$ and with the complementary probability $(1-p_0)$ it follows the distribution function $F_2(x)$:

$$P\{z_k < x\} = p_0 F_1(x) + (1 - p_0) F_2(x). \tag{2.19}$$

The passage from an exponential law to a power law can thus be expected, if positive feedback (and so, synergetic effects) can occur for larger events.

2.4 The Main Results

In the first chapter empirical data characterizing both the physical parameters of disasters and disaster loss values were examined, and it has been shown that these empirical distributions can be well fitted by a few typical distribution laws: by the normal Gaussian distribution, by the Boltzmann exponential distribution, and by the Pareto power-law distribution, as well as by their combinations. A classification of different kinds of disaster into three types was proposed. This classification takes into account an essential distinction in the distribution laws typical of different types of natural disasters. This approach was developed in the present chapter by examining the conditions needed for the occurrence of these typical distribution laws. Some models of stochastic systems generating exponential and power-law distributions are demonstrated. Simple additive (cumulative) and multiplicative models were proposed. The multiplicative model can lead under certain conditions to the power-law distribution with a heavy tail.

We have demonstrated one common feature inherent in all models generating heavy-tailed distributions: heavy tails can occur when a product of factors surpassing unity appears in the model. This cause is rather typical. A combined synergetic effect of such factors can considerably increase the total effect. The avalanche-like processes of natural disasters can be caused by positive feedbacks.

For a more complex case when the empirical distribution shows the features relevant both to the exponential and to the power-law distribution we proposed a mixture of models.

One should note, however, that the present discussion on power-law distributions is incomplete, because such distributions frequently occur in the critical-state phenomena and take place in the self-organized criticality model [SS1, Sor3].

This approach is undoubtedly fairly simplified, so it satisfies the first requirement of Einstein's epigraph to this chapter. Whether this approach also satisfies the second requirement will be left for the readers to judge.

Chapter 3
Nonparametric Methods in the Study of Distributions

The two first chapters were devoted to a classification of disasters and their parametric quantification. However, no statistical study of disasters will be complete without some nonparametric approach to the study of the relevant distributions. The examples in the first chapter show that for many types of disasters there exists no universal method to parameterize a distribution. For most types of disasters, in particular, for earthquakes, the tail distribution (that for the topmost range of rare events) is quite controversial. So, when there is no generally accepted parameterization, it seems reasonable to use nonparametric estimates. Such estimates are free of possibly inadequate assumptions about parameterization of empirical distributions. Of course, one should pay for this advantage: nonparametric methods are generally less efficient than parametric ones.

We set forth in this chapter several methods to use in nonparametric estimation of distributions. The stress is on the study of tail distributions. We derive confidence limits for the distribution of event size and for the Poisson intensity of a sequence of events. A statistical estimate of the probability of exceedance above a record event (maximum magnitude in the catalog) in a future time period is provided, as well as a confidence interval for this probability. The distribution of the time interval until the next record has been derived.

3.1 Application to Earthquake Catalogs

A good example of application of nonparametric methods is provided by large earthquakes. The distribution of earthquake magnitudes in the range of small and moderate values and in large spatio-temporal volumes is known to obey the fundamental Gutenberg–Richter law, i.e., the logarithmically linear dependence of the number $N(m)$ of earthquakes whose magnitudes exceed a given magnitude m. Because the logarithm of earthquake energy depends linearly on magnitude, the energy itself has a power-law distribution (the Pareto distribution). However, deviations from this general dependence are observed in detailed seismic studies

V. Pisarenko and M. Rodkin, *Heavy-Tailed Distributions in Disaster Analysis*, Advances in Natural and Technological Hazards Research 30,
DOI 10.1007/978-90-481-9171-0_3, © Springer Science+Business Media B.V. 2010

and are most prominent at large magnitudes. Various authors have proposed modifications of the Gutenberg–Richter law in the range of large magnitudes [GR, MP, CFL, P2, We, MKDN, K4, K5, K8, K9, KS2, PS2].

Since large events have been comparatively few during the period of instrumental seismology, even on a global scale, all these modifications involve inevitable uncertainties, and there is no agreement among seismologists regarding the behavior pattern of the distribution law of magnitudes (or other characteristics of earthquake size, such as the scalar seismic moment) in the range of the largest possible values. The best-known modification [K4] modulates the exponential energy distribution by an exponential factor that decreases with increasing argument, which yields a gamma, rather than power-law, distribution describing rather adequately the observed downward deviations from the logarithmically linear Gutenberg–Richter relation.

Because of the aforementioned insufficient number of large earthquakes, results from application of any parametric modification substantially depend on a priori assumptions about the families of functions to be used for the description of the energy distribution in the range of large values. Therefore, it is of interest to try to describe this distribution without any a priori assumptions as to the distribution shape. Such an attempt is made in the present work on the basis of nonparametric methods, in particular, the theory of records (see [Ga, EKM]. Although the majority of these methods are known in mathematical statistics, they have not found as yet adequate applications in seismology.

Some approaches to the analysis of seismic hazard using the theory of records are proposed in Dargahi-Noubary [Dar]. However, the main results are obtained in this work for the unrestricted Gutenberg–Richter law, while the most interesting problems are related precisely to the estimation of deviations of an empirical distribution from this law. Moreover, the problem of confidence intervals was not considered for the parameters estimated (we pay special attention to this question because of its particular importance in applied problems).

Denote as $X(t_1), \ldots, X(t_n)$ a sequence of measured earthquake values (magnitude or seismic moment) reported in a regional or global catalog ($t_1, \ldots, t_n$ are the corresponding time moments). We are interested in the size distribution $X_k = X(t_k)$ in the range of large values (the tail). No parametric forms of the distribution function are used within the framework of the nonparametric approach, and our sole assumption is as follows. The occurrence times of the events form a stationary Poisson process. More specifically, the occurrence times form independent stationary Poisson processes for any two nonintersecting size intervals $Y_1 < X \leq Y_2$ and $Y_3 < X \leq Y_4$. The sequence $X(t_1), \ldots, X(t_n)$ can be considered as a marked point Poisson process [DV].

It should be noted that this assumption contradicts the well-known seismological concepts, namely, the presence of aftershocks, foreshocks, and earthquake swarms, whose sequences of occurrence times are obviously statistically interdependent. However, since we are mainly interested in the behavior of distribution tails, smaller events (including aftershocks and foreshocks) are insignificant here. Moreover, we assume that aftershocks have been removed from the catalog using some well-known methods (e.g., see [MD]). Certainly, it is conceivable that pairs of relatively large events and other sets of events that cannot be considered as fully independent will remain in the catalog after application of such rejection procedures. However, the study of the statistics of such phenomena is hampered by the small number of these and, in our opinion, the disregard of such rare dependent events cannot affect the results presented below. We believe that the assumption that the process is Poissonian is sufficiently justified for main shocks (see [PSSR]).

Our assumption implies that, given a fixed number of observations n, the sample $X_1, \ldots, X_n$ can be considered as a sample of n independent random variables having a common distribution function. Thus, a link is established between the temporal sequence of earthquakes (a marked point process) and a sample of n independent identically distributed observations with a common distribution function. As the observation time covered by the catalog increases, the random Poisson number n tends to infinity with probability one, and asymptotic statistical inferences drawn from a sample of n independent observations become valid for a marked point process.

3.2 Estimates of the Lower and Upper Bounds for the Tail of a Distribution Function

Suppose we have a sample $X_1, \ldots, X_n$ of independent random variables with a common continuous distribution function $F(x)$. We have to estimate the probability $1 - F(x)$ from above and from below (note that the entire distribution function can be analyzed in the same way).

We fix x and denote the tail probability by p:

$$p = 1 - F(x) \tag{3.1}$$

Let the number of sample values exceeding x be m. The random variable m obeys the binomial distribution

$$P\{m = k\} = C_n^k p^k (1-p)^{n-k}, \quad k = 0, 1, \ldots n. \tag{3.2}$$

The sum of probabilities (3.2) is given by the formula [BS]

$$P_r = P\{m \le r\} = \begin{cases} \sum\limits_{k=0}^{r} C_n^k p^k (1-p)^{n-k} = 1 - B_p(r+1, n-r), & r < n; \\ 1, & r = n. \end{cases} \tag{3.3}$$

where $B_p(r+1, n-r)$ is the incomplete (normalized) B-function. Now consider the solution $\pi(m/\varepsilon, n)$ of the equation

$$B_\pi(m, n-m+1) = \varepsilon/2, \tag{3.4}$$

(when $m = 0$ the solution will be $\pi = 0$). This solution depends on $m = m(x)$ and is therefore a random variable. It is easy to show that $P\{p \geq \pi\} = 1-\varepsilon/2$ [BS]. Thus, π is a lower bound with confidence level $1 - \varepsilon/2$:

$$P\{p \geq \pi\} = 1 - \varepsilon/2 \tag{3.5}$$

The equation for the upper bound Π with confidence level $1 - \varepsilon_1/2$ is obtained similarly:

$$B_{\Pi}(m+1, n-m) = 1 - \varepsilon_1/2, \quad m < n; \tag{3.6}$$

$\Pi = 1$ should be taken at the point $m = n$:

The confidence interval with confidence level $(1 - \varepsilon/2 - \varepsilon_1/2)$ (we assume that $\varepsilon + \varepsilon_1 < 2$) is

$$P\{\pi \leq p \leq \Pi\} = 1 - \varepsilon/2 - \varepsilon_1/2. \tag{3.7}$$

If $\varepsilon = \varepsilon_1$, the confidence interval (3.7) has the confidence level $1 - \varepsilon$.

Thus, the determination of the upper and lower confidence bounds for the tail probability p reduces to the solution of Eqs. 3.4 and 3.6, which can be done with the help of a standard program. The resulting estimates are nonparametric, i.e., do not depend on any assumptions about the distribution law (e.g. the Gutenberg–Richter law or modifications of this).

The confidence interval (3.7) is applicable with any m, n ($n = 1, 2, \ldots; 0 \leq \mathrm{m} \leq \mathrm{n}$) to an arbitrary distribution function. For example, the interval (3.7) can be used to estimate the tail of the conditional distribution of the energy of large earthquakes exceeding a certain threshold and to construct confidence intervals (although they can be very wide) for tails consisting of only a few earthquakes.

To illustrate the application of the approaches described above, we used two samples from the world Harvard catalog for the Sunda (Sumatra) and Pakistan regions over the period 1976–2005. We selected events with scalar seismic moments above 3×10^{24} *dyne cm*, which made the plot a straight line. Shallow earthquakes with depths $h \leq 70$ km were used. The Pakistan subcatalog contains 57 earthquakes with maximum magnitude $m_w = 7.6$. The Sunda subcatalog contains 166 earthquakes and their maximum magnitude is $m_w = 9.0$. Empirical normalized frequency-magnitude plots are shown for these regions in Fig. 3.1.

Figure 3.2 presents 90% confidence limits for the tail of the seismic moment distribution in the Sunda region. The confidence intervals are rather wide in the range of high magnitudes, but fortunately these were derived without any initial assumption as to the tail shape.

To obtain a confidence interval for the unknown probability p, we can use the limit distribution of the estimate p = m/n provided by the Moivre–Laplace theorem, which states that the standard Gaussian distribution is the limit distribution of the random variable $(p - m/n)/(n{\cdot}p(1 - p))^{1/2}$. To find the upper confidence bound with

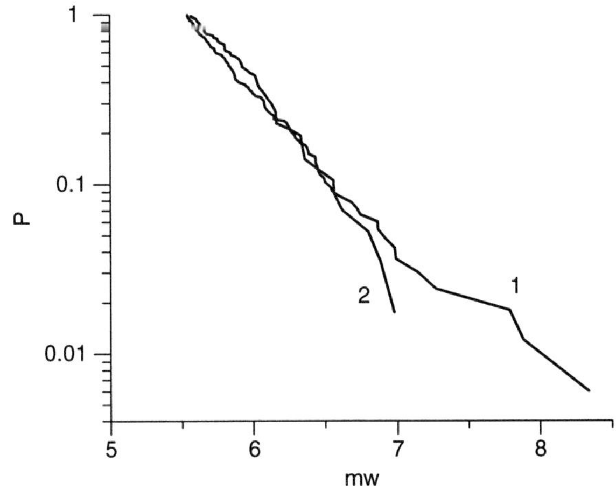

Fig. 3.1 Complementary sample distribution functions: (1) Sunda, (2) Pakistan

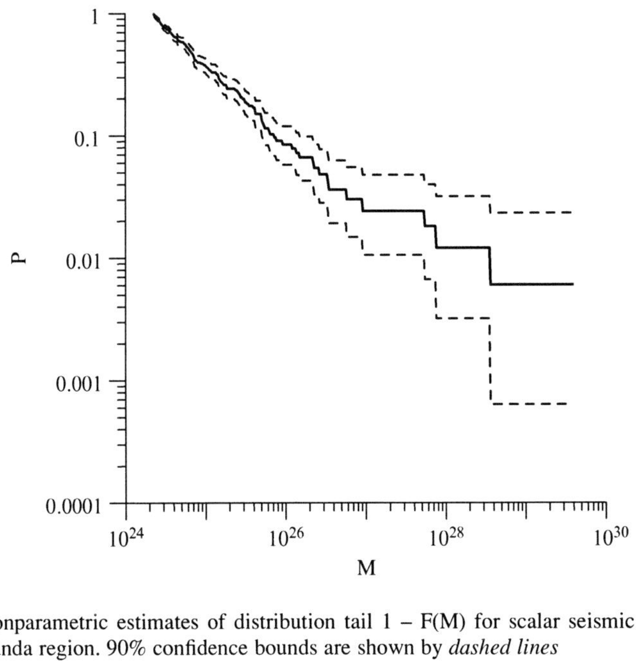

Fig. 3.2 Nonparametric estimates of distribution tail 1 – F(M) for scalar seismic moment M, dyne-cm. Sunda region. 90% confidence bounds are shown by *dashed lines*

confidence level $(1 - \varepsilon/2)$ we take the Gaussian quantile $g = g(\varepsilon)$ at level $(1 - \varepsilon/2)$ and assume that the equality

$$P\{(p - m/n)/(n \cdot p(1-p))^{1/2} \leq g\} = 1 - \varepsilon/2. \tag{3.8}$$

is true. Solving the equation

$$(p - m/n)/(n \cdot p(1-p))^{1/2} = g$$

for p, we obtain two different real roots Q and q,

$$Q = \left(\hat{p} + g^2/2n + \left((g^2/2n)^2 + \hat{p}(1-\hat{p})g^2/n\right)^{1/2}\right) \Big/ \left(1 + g^2/n\right); \quad (3.9)$$

$$q = \left(\hat{p} + g^2/2n - \left((g^2/2n)^2 + \hat{p}(1-\hat{p})g^2/n\right)^{1/2}\right) \Big/ \left(1 + g^2/n\right); \quad (3.10)$$

$$\hat{p} = m/n.$$

These are, respectively, the upper and lower confidence limits for p at the confidence level $(1 - \varepsilon)$

$$P\{q \le p \le Q\} = 1 - \varepsilon. \quad (3.11)$$

Note that (3.11) is true only asymptotically where the standard Gaussian distribution becomes a good approximation for $(p - m/n)/(n \cdot p(1 - p))^{1/2}$. Numerical examples show that equality (3.11) can be used for n > 20; the correct confidence interval (π, Π) for $n < 20$ is noticeably larger than the interval (q, Q), and the ratio of lengths of these intervals can reach 1.5 or more. Thus, the exact expressions (3.4)–(3.7) are preferable at smaller n.

3.3 Confidence Intervals for the Intensity of a Poisson Process

Let $\lambda(M) \cdot \Delta$ denote the intensity of a Poisson sequence of earthquakes in the size interval $(M, M + \Delta)$. Since the sequences in nonoverlapping size intervals are assumed to be independent, the intensity in an interval (M_1, M_2) is expressed by the integral

$$\int_{M_1}^{M_2} \lambda(M) dM. \quad (3.12)$$

The distribution density of earthquake size $f(x) = F'(x)$ in the interval (m, M_{max}) coincides with the normalized intensity function $\lambda(M)$:

$$f(x) = \lambda(x) \Big/ \int_{m}^{M\,\mathrm{max}} \lambda(M) dM. \quad (3.13)$$

Thus, the density $f(x)$ expresses the normalized intensity of a sequence of events of various sizes.

The number ν_T of events that occurred during a time T in an arbitrary size interval (M', M'') is a Poisson random variable with the parameter Λ:

$$\Lambda = T \int_{M'}^{M''} \lambda(M) dM \tag{3.14}$$

An exact confidence interval $(\underline{\Lambda}, \overline{\Lambda})$ can be obtained for the parameter Λ [BS]:

$$P\{\underline{\Lambda} < \Lambda < \overline{\Lambda}\} = 1 - \varepsilon, \tag{3.15}$$

where $\underline{\Lambda}, \overline{\Lambda}$ are the solutions to the equations

$$\chi^2_{2\nu_T}(2\underline{\Lambda}) = \varepsilon/2; \quad \chi^2_{2+2\nu_T}(2\overline{\Lambda}) = 1 - \varepsilon/2; \tag{3.16}$$

and $\chi^2_k(x)$ denotes the chi-square distribution with two degrees of freedom. Hence the confidence interval for the intensity

$$\int_{M'}^{M''} \lambda(M) dM.$$

is $(\underline{\Lambda}/\, T, \overline{\Lambda}/\, \mathrm{T})$.

We use (3.15) and (3.16) to estimate the probability that at least one earthquake of a size at least M occurs during a (future) time interval $(t, t+T)$. We denote as λ_M 91/*yr*) the intensity of the sequence of main shocks of size no less than M. The probability $P_{M,T}$ that at least one such earthquake will occur in the future time interval $(t, t + T)$ is

$$P_{M,T} = 1 - exp(-T\lambda_M). \tag{3.17}$$

Let n_M shocks of magnitude no less than M have occurred in a region in a past time interval $(t-\tau, t)$. The random variable n_M has a Poisson distribution with the parameter $\tau \cdot \lambda_{\mathrm{M}}$. The lower and upper confidence limits $\underline{\Lambda}, \overline{\Lambda}$ at confidence level $(1 - \varepsilon/2)$ can be determined for this parameter from Eq. 3.16.

If we consider the number of earthquakes of a size no less than M in a future time interval $(t, t + T)$, the confidence interval for the corresponding Poisson parameter at the confidence level $(1 - \varepsilon)$ has the form $(T\, \underline{\Lambda}/\tau;\ T\, \overline{\Lambda}/\tau)$. Thus, the confidence interval at the confidence level $(1 - \varepsilon)$ for $P_{\mathrm{M},T}$ is, in accordance with (3.17),

$$\left(1 - exp(-T\underline{\Lambda}/\tau);\quad 1 - exp\left(-T\overline{\Lambda}/\tau\right)\right). \tag{3.18}$$

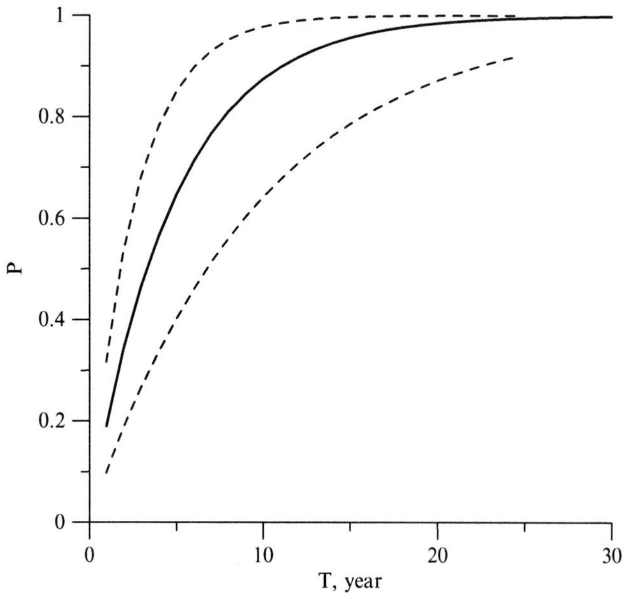

Fig. 3.3 Nonparametric estimate of probability of occurrence for at least one earthquake exceeding $M_W = 7.0$, Sunda region (*solid line*). 90% confidence bounds (*dashed lines*)

Figure 3.3 presents an example illustrating how the probability that more than one earthquake of a magnitude no less than 7.0 will occur in the Sunda region in a future time interval T depends on the length of this interval.

An exact confidence interval can also be obtained for the ratio ρ

$$\rho = \int_{M'}^{M''} \lambda(M)dM \Big/ \int_{M'''}^{M''''} \lambda(M)dM \tag{3.19}$$

of the intensities Λ_1 and Λ_2 of the sequences of events in two arbitrary intervals (M', M'') and (M''', M'''') (BS). If the numbers v_1 and v_2 of events observed over a time τ in these intervals are known, then

$$P\left\{\underline{\rho} < \rho < \overline{\rho}\right\} = 1 - \varepsilon, \quad \begin{array}{ll} \underline{\rho} = \begin{cases} 0; & v_1 = 0 \\ (v_1/(v_2+1))/F(1-\varepsilon/2, 2v_2+2, 2v_1)) & v_1 > 0; \end{cases} \\ \overline{\rho} = \begin{cases} \infty & v_2 = 0; \\ ((v_1+1)/v_2)F(1-\varepsilon/2, 2v_1+2, 2v_2); & v_2 > 0; \end{cases} \end{array} \tag{3.20}$$

Here $F(x,m_1,m_2)$ denotes the Fisher distribution with the (m_1,m_2) degrees of freedom. As an illustration of formula (3.20), Fig. 3.4 shows the ratio ρ of intensities

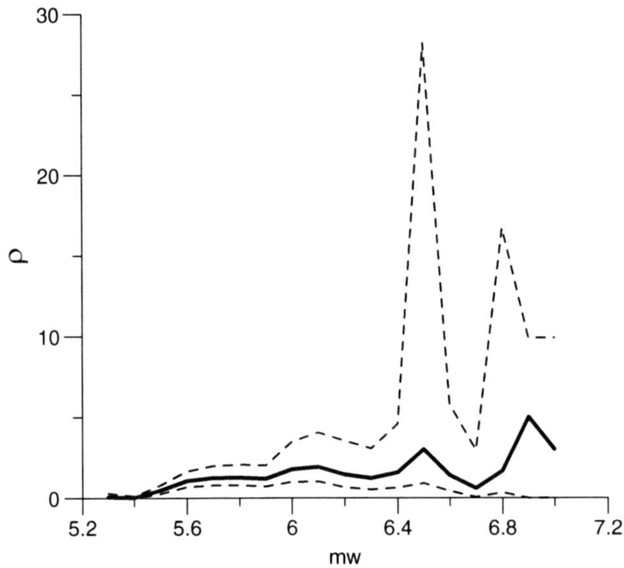

Fig. 3.4 Ratio of intensities ρ in adjacent 0.1 magnitude intervals, Pakistan (*solid line*). 68% confidence bounds (*dashed lines*)

calculated at 68% confidence level in adjacent magnitude intervals for the Pakistan region, each interval being equal in width to 0.1 of m_W unit. The ratio of intensities is seen to be nearly constant at $m_W < 6.2$, which agrees with the Gutenberg–Richter relation (self-similarity of seismicity), but peaks arise in the range of large magnitudes near half-integer values, apparently due to a tendency of rounding off to half-integer values involved in the processing of raw catalog data. This plot illustrates the sensitivity of the method of estimating the ratio of intensities.

A confidence interval for any monotone increasing function $G(\rho)$ can easily be obtained from (3.20):

$$P\left\{G(\underline{\rho}) < G(\rho) < G(\overline{\rho})\right\} = 1 - \varepsilon. \tag{3.21}$$

3.4 Probability of Exceeding a Past Record in a Future Time Interval

Gumbel [Gu1] solved the following problem using combinatorial considerations. A sample of independent identically distributed random variables $X_1, \ldots, X_n$ obeying a continuous distribution function is given. We denote

$$H_n = max(X_1, \ldots, X_n).$$

An additional independent sample $Y_1, \ldots, Y_r$ with the same distribution function is drawn. What is the probability $P_n^m(r)$ that m of these r observations ($0 \le m \le r$) exceed H_n (m exceedances over the maximum H_n will occur)? The answer does not depend on the distribution function:

$$P_n^m(r) = C_{n+r-m-1}^{n-1} / C_{n+r}^{n} = (n + r - m - 1)!r!/(r - m)!(n + r)! \tag{3.22}$$

We extend this formula to the case where the additional sample $Y_1, \ldots, Y_r$ consists of a random number of observations r, provided that r is the number of events of the Poisson process in a time interval T and the associated intensity is Λ (consequently, the Poisson parameter for the random number r is $\Lambda \cdot T$). The Y and r are assumed to be independent. Considering conditional probabilities that m exceedances are observed at $r = 0, 1, 2, \ldots$, we obtain the desired probability $P_n^m(\Lambda T)$ from the formula for the total probability:

$$P_n^m(\Lambda T) = \sum_{r=0}^{\infty} P_n^m(r) \exp(-\Lambda T)(\Lambda T)^r / r! \tag{3.23}$$

The probability $P_n^0(\Lambda T)$ at $m = 0$ (the probability that there are no exceedances) or the complementary probability (the probability that one or more events will exceed the maximum H_n) are those of most interest from the practical standpoint. We have from (3.23):

$$P_n^{\text{o}}(\Lambda T) = \exp(-\Lambda T) \sum_{r=0}^{\infty} (\Lambda T)^r / ((1 + r/n) r!) \tag{3.24}$$

Unfortunately, the probability (3.24) cannot apparently be expressed explicitly through special functions, but can easily be estimated numerically. The probabilities that there are no exceedances over the maximum H_n can be calculated from (3.24) for a given set of T values that are of interest for applications.

Using the confidence interval (3.15) for the intensity Λ and the monotone dependence of (3.24) on Λ, we can determine the confidence interval for $P_n^0(\Lambda T)$:

$$P\{P_n^0(\underline{\Lambda} T) < P_n^0(\Lambda T) < P_n^0(\bar{\Lambda} T)\} = 1 - \varepsilon. \tag{3.25}$$

The probability that the maximum magnitude $m_W = 7.6$, as observed in the interval 1976–2005 (29.25 years) for the Pakistan region, will be exceeded in a future interval of T years ($1 \le T \le 50$) is shown in Fig. 3.5. The 90% confidence interval is also shown.

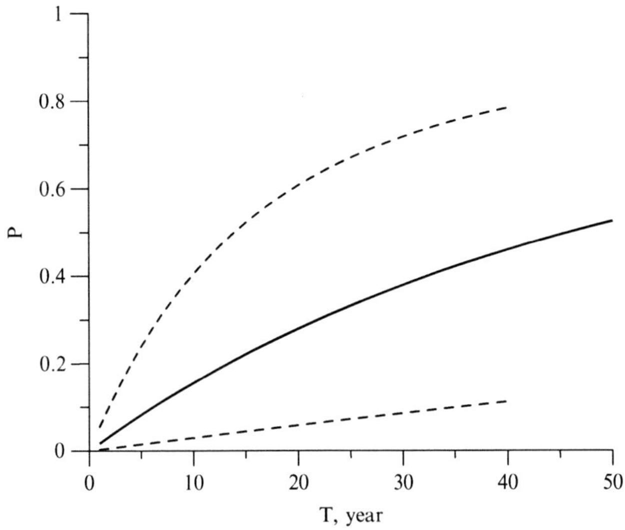

Fig. 3.5 Probability of exceeding the past maximum magnitude $M_W = 7.6$ (1976–2005) as a function of future time span T (*solid line*), the 90% confidence interval (*dashed lines*)

3.5 Distribution of the Time to the Nearest Event Exceeding the Past Maximum

We consider a catalog $X_1, \ldots, X_n$ as a sample of independent identically distributed random variables obeying a certain continuous distribution function and observed until the time moment $t = 0$. As before, we denote

$$H_n = max(X_1, \ldots, X_n). \tag{3.26}$$

Let v be a random number of observations preceding the first exceedance over H_n:

$$X_{n+1}, \ldots, X_{n+v}; \quad X_{n+k} < H_n, \quad k < v; \quad X_{n+v} > H_n.$$

We are to find the probabilities $P\{v = j\}$, $j = 1, \ldots$. Because any monotone transformation of the $\{X\}$ does not change these probabilities, we can assume that the $\{X\}$ are distributed uniformly on the unit interval. The distribution function of H_n is then x^n and we have, according to the formula for the total probability,

$$P\{v = j\} = \int_0^1 P\{v = j | H_n = x\} dx^n = n \int_0^1 x^{n+j-2}(1-x)dx = n/(n+j)(n+j-1). \tag{3.27}$$

We denote as τ_i the intervals between successive events. The time interval T until the first exceedance is thus equal to

$$T = \tau_1 + \tau_2 + \cdots + \tau_v.$$

The unconditional distribution function of T is found from the formula for total probability:

$$P\{T < x\} = \sum_{j=1}^{\infty} P\{\tau_1 + \tau_2 + \cdots + \tau_j < x\} P\{v = j\}. \tag{3.28}$$

The random variables $2\Lambda \cdot \tau_i$ have a chi-square distribution with two degrees of freedom. Consequently, the sum $2\Lambda \cdot (\tau_1 + \tau_2 + \cdots + \tau_j)$ has a chi-square distribution with $2j$ degrees of freedom. We obtain

$$\begin{aligned} P\{\tau_1 + \tau_2 + \cdots + \tau_j < x\} &= P\{2\Lambda(\tau_1 + \tau_2 + \cdots + \tau_j) < 2\Lambda x\} \\ &= \chi^2_{2j}(2\Lambda x), \end{aligned} \tag{3.29}$$

where $\chi^2_{2j}(Z)$ denotes the value of the chi-square distribution function with $2j$ degrees of freedom at the point z:

$$\chi^2_{2j}(z) = 1/\left(2^j(j-1)!\right) \int_0^z u^{j-1} \exp(-u/2) du.$$

Substituting (3.29) into (3.28), we find

$$P\{T < x\} = n \sum_{j=1}^{\infty} \chi^2_{2j}(2\Lambda x)/(n+j-1)(n+j). \tag{3.30}$$

Formula (3.30) can be reduced to a recurrent form convenient in practice. We define functions $H_k(u)$

$$H_k(x) = \begin{matrix} 1 - exp(-\Lambda x); & k = 1; \\ x^{k-1} - 2(k-1)H_{k-1}(x); & k > 1, \end{matrix} \tag{3.31}$$

and functions $G_k(x)$

$$G_k(x) = \begin{matrix} C + log(\Lambda x) + expint(\Lambda x); & k = 1; \\ G_{k-1}(x) - H_k(x)/(k-1)(2\Lambda x); & k > 1, \end{matrix} \tag{3.32}$$

where $C = 0.577216\ldots$ is the Euler constant. One can readily show that

$$P\{T < x\} = n(G_n(x) - G_{n+1}(x)). \tag{3.33}$$

The standard special function *expint*(z) (the exponential integral) has the form

$$expint(z) = \int_{z}^{\infty} (e^{-t}/t)dt, \quad z > 0;$$

Formula (3.33) is suitable for numerical calculation of the distribution function for T or lgT. Setting $x = 10^y$, we obtain

$$P\{lg(T) < y\} = n(G_n(10^y) - G_{n+1}(10^y)).$$

We denote the distribution function (3.30) by $\Psi(x,\Lambda)$. This function depends on the parameter Λ. Because this parameter is unknown in practice and should be estimated from a limited catalog, its estimate is subject to random error. If this estimate is used in (3.30), $\Psi(x,\Lambda)$ will also involve a random error. We present the confidence interval for the true value of $\Psi(x,\Lambda)$ making use of the confidence interval for Λ (3.15) and the fact that $\Psi(x,\Lambda)$ is a monotone function of Λ at any x:

$$P\left\{\Psi(x,\underline{\Lambda}) < \Psi(x,\Lambda) < \Psi(x,\overline{\Lambda})\right\} = 1 - \varepsilon. \tag{3.34}$$

Note that the use of the same segment of a catalog for estimating the parameter Λ and the number of records in (3.30) is not quite correct due to their probable interdependence. To avoid this difficulty, we propose the use of different parts of the catalog for obtaining these estimates. For example, Λ can be estimated from the events that occurred in even years of the catalog, while uneven years are used for calculating the sample maximum. Of course, the catalog length for each of these estimates will be twice as small, but these estimates will be obtained from non-overlapping data.

We illustrate the method described above using as an example the Pakistan region, where 361 events with magnitudes $4.4 \le m_W \le 7.6$ occurred from July 28, 1976 through October 19, 2005. Four of these events had magnitudes higher than 6.5:

December 30, 1983, $m_W = 6.7$;
July 29, 1985, $m_W = 6.7$;
October 19, 1991, $m_W = 6.5$;
October 8, 2005, $m_W = 6.7$.

The estimate for the intensity of events with magnitudes no less than 6.5 is $\Lambda = 4/29.85 = 0.134$ events per year. The distribution function of lgT and the corresponding (normalized to maximum) density are shown in Fig. 3.6. The sample quantiles have the following values:

$$\begin{aligned} &Q_{0.16}(lgT) = 0.842; \quad Q_{0.16}(T) = 6.95 \; year \\ &Q_{0.50}(lgT) = 1.534; \quad Q_{0.50}(T) = 34.18 \; year; \quad median; \\ &Q_{0.84}(lgT) = 2.217; \quad Q_{0.84}(T) = 164.82 \; year. \end{aligned}$$

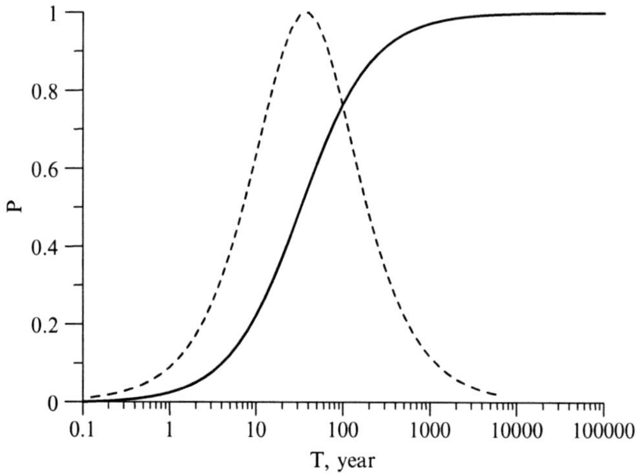

Fig. 3.6 Distribution function (*solid line*) and the probability density normalized by the maximum (*dashed line*) of the time T until the next record exceeding the past record $M_W = 7.6$, Pakistan

We chose quantiles at these levels because, in the case of the normal law, the level 0.50 (median) corresponds to the mean m and the levels 0.16 and 0.84 correspond, respectively, to $m - \sigma$ and $m + \sigma$, where σ is the standard deviation of the normal law.

We see that the distribution density of lgT is very close to the Gaussian law; i.e., T has a nearly lognormal distribution. Ga proved that lgT does converge in the limit to the normal law as the sample size increases, and $T \sim exp(r_n)$, where r_n is the number of records in a sample of size n; however, for finite n, the tail of the distribution of lgT is found to be somewhat heavier than the tail of the Gaussian distribution.

3.6 Main Results

We described some methods of statistical analysis to study the tail of the earthquake energy distribution that involve no assumptions about the form of the distribution law. Although the majority of the methods proposed here are well known in mathematical statistics, their application to seismology is apparently novel.

Naturally, nonparametric methods of statistical analysis are less effective than parametric methods, but they are free of a priori (rather controversial) assumptions about the form of the distribution law in the range of large values. Some loss in efficiency is the payment for a greater generality. In spite of their lower efficiency, the proposed nonparametric methods, in our opinion, are of interest for applications. Thus, the distribution of the random time T to the next magnitude record in a region can be useful in the analysis of regional seismicity (Fig. 3.6). The distribution of T is very close to the lognormal type and, consequently, has a rather heavy tail. Therefore, it is better from the practical standpoint to estimate $logT$ and its quantiles.

The dependence of the probability of exceedance over a preceding record within a given future time interval on the length of the interval is also of interest (Fig. 3.5). Plots similar to the one shown in Fig. 3.5 can be used to derive constraints on the probability of recurrence of events comparable with a past maximum event in a given region and to estimate the uncertainty of the resulting probability. Another example illustrating the practical significance of the probability of exceedance over a known record level is for the design of dam height for flood protection [Ha, EKM]; in this case, the application of the nonparametric approach can prove necessary because no generally accepted parameterization of the distribution law exists for the flood-related water level heights.

Strictly speaking, the methods described in this paper are applicable only to random variables with a continuous distribution and, therefore, they are formally inapplicable to catalogs with digitized magnitudes (this digitization is presently performed at a step of 0.1). The step 0.1 is very small, and this accuracy is usually sufficient for any practical purposes. Nevertheless, sometimes identical values of record magnitudes can occur. This fact prevents a straightforward use of our methods. However, this difficulty can be overcome with the help of the following procedure. Magnitudes can be randomly spread by adding random variables uniformly distributed in the interval $(-0.05; 0.05)$ or in a smaller interval and subsequently averaging the final estimates of several such spreading procedures. The series of magnitudes spread in this way will not contain repeated values and the estimation methods described above can be safely used. Of course, details of such a spreading procedure deserve a more careful study.

Chapter 4
Nonlinear and Linear Growth of Cumulative Effects of Natural Disasters

4.1 Nonlinear Growth of Cumulative Effects in a Stationary Model with the Power (Pareto) Distribution

As was shown earlier, the economic losses and casualties from natural disasters are often fitted by self-similar power-law distributions with heavy tails. If the exponent β of such a distribution does not exceed unity, $\beta \leq 1$, then the mathematical expectation is infinite. In this case, the standard statistical tools, such as sample mean and sample standard deviation, are highly unstable (see Table 1.1). The increase of sample size, in contrast to the ordinary case of finite expectation, does not enhance the accuracy of the sample mean. The large statistical scatter of sample means (which are widely used until today, see e.g., [O1, O2, O3]) makes them inadequate for practical use in problems of loss reduction. We discuss below some approaches to the reliable estimation and prediction of total effects for the cases in which the standard statistical tools are inefficient.

4.1.1 The Existence of a Nonlinear Growth of Cumulative Effects in a Stationary Model with the Pareto Distribution

We shall demonstrate a nonlinear growth of cumulative effects in a stationary model. We consider a stationary sequence of independent, identically distributed random variables X_k with the Pareto distribution function $F(x)$ and density $f(x)$:

$$F(x) = 1 - 1/x^{\beta}, \quad x \geqslant 1; \tag{4.1}$$

$$f(x) = \beta/x^{1+\beta}, \quad x \geqslant 1; \tag{4.2}$$

We can interpret X as random loss in a time series of some natural disaster. Let us assume for simplicity that the intensity of events λ is unity (say, one event per

V. Pisarenko and M. Rodkin, *Heavy-Tailed Distributions in Disaster Analysis*, Advances in Natural and Technological Hazards Research 30,
DOI 10.1007/978-90-481-9171-0_4,

year). We are going to show that the total loss for n years $\sum_n$ behaves quite differently, depending on the fact whether the index β is greater than unity or not. Let us compare two particular values: $\beta = 3$ and $\beta = 0.7$. The mean m and the variance *var* of X are easily found, if $\beta > 2$:

$$m = \beta/(\beta - 1); \quad \text{var} = \beta/[(\beta - 1)^2 \cdot (\beta - 2)].$$

In the former case $m = 1.5$; $var = 0.75$. Both characteristics are finite. Thus, by virtue of the Central Limit Theorem in probability theory, the loss for 30 years $\sum_{30}$ is fitted to good accuracy by a normal variable with mean $30{\cdot}m = 45$ and standard deviation $(30{\cdot}var)^{1/2} = 4.74$:

$$\sum\nolimits_{30} \cong 45 + 4.74 \cdot \zeta,$$

where ζ is a standard normal variable. Thus, in this case, as was to be expected, the total effect $\sum_{30}$ is fairly accurately equal to the 30-fold yearly loss, i.e., the total loss grows in a linear manner with time.

In the latter case ($\beta = 0.7$), the mathematical expectation is infinite, so we cannot use it to describe cumulative losses. Instead, we use the median μ of the distribution (4.1), i.e., the root of equation $F(x) = 0.5$. It is easily found that $\mu = 2^{1/\beta} = 2.69$. It is not easy to derive the distribution of the total effect for the Pareto distribution with infinite expectation. That is why we estimate the total effect $\sum_{30}$ from below by means of its maximum term, which is sure to be below $\sum_{30}$. As a statistical characteristic of the maximum we take its median μ_{30}. The distribution function of the maximum equals the distribution function $F(x)$ of a single event raised to the power of 30, and the median μ_{30} is found as the root of $F^{30}(x) = (1 - 1/x^{\beta})^{30} = 1/2$. We find $\mu_{30} = (1 - 2^{-1/30})^{-1/\beta} \cong 221$. In contrast to the former case, the total loss for 30 years is about $221/2.69 = 82$ times the typical yearly loss! The growth of the total effect is thus nonlinear. For a time interval of 60 years a similar evaluation gives $\mu_{60} = (1 - 2^{-1/60})^{-1/\beta} \cong 590$, which is 220 times the typical yearly loss!

These comparisons show that the nonlinear growth of the total effect can be very considerable for heavy-tailed distributions such as the Pareto law with $\beta < 1$. The cause of this nonlinear growth consists in the fact that the probability of a giant event that can produce an effect comparable with the total effect of all other events, or even exceeding it, is growing with time. We stress that this nonlinear growth with time was obtained for a stationary model (a stationary time series of independent, identically distributed random variables). We should note, however, that in real situations such a nonlinear growth can be observed only for a limited time period, even if the distribution of an individual effect is well fitted by the Pareto law with $\beta < 1$. The reason is that such a fit cannot well continue satisfactorily to infinity: there are no infinite effects in natural phenomena. The size of a natural disaster is in principle limited by the finite dimensions of our planet. The transition from nonlinear to linear mode of growth required by this limitation will be studied below for the case of earthquake losses and seismicity behavior. But such a transition is also typical of other natural disasters.

4.1.2 The Evaluation of the Maximum Individual Loss

We used in the preceding section the median as a statistical characteristic of the typical size of an event obeying a heavy-tailed distribution (e.g., the Pareto distribution with $\beta < 1$). We could not use the sample mean, since its theoretical analogue is infinite, and it diverges with increasing n. In contrast to this behavior, the sample median m^* tends to its theoretical analogue m as the sample size increases, its normalized deviation

$$n^{1/2} \cdot (m^* - m)/s$$

converging to the standard Gaussian random variable [Cr]. Here s is the asymptotic standard deviation:

$$s = [2 \cdot f(m)]^{-1}, \tag{4.3}$$

where $f(x)$ is the probability density of an individual effect. For practical uses, when $f(x)$ is unknown, it is necessary to obtain a satisfactory estimate of this density, e.g., by the kernel method [Har].

We now set forth in detail the method for estimating the maximum loss caused by an event that will occur in a future time interval τ. The sequence of events (disasters) is assumed to be a Poisson stationary process with intensity λ events per year. The intensity λ is estimated from the catalog consisting of n events by the customary procedure:

$$\hat{\lambda} = n/T, \tag{4.4}$$

where T is the time interval covered by the catalog. The Poissonian random variable n has the variance λT. Thus, the std of (4.4) is

$$std(\hat{\lambda}) = (\lambda T)^{1/2}/T = (\lambda/T)^{1/2}.$$

Since λ is unknown, we can replace it with its estimate (4.4):

$$std(\hat{\lambda}) \cong n^{1/2}/T.$$

For a Poissonian sequence of events with the distribution function of an individual event $F(x)$, the distribution function of the maximum event for a time interval τ, denoted as $\Phi_\tau(x)$, is well known (see, e.g., [P3]):

$$\Phi_\tau(x) = [exp(\lambda\tau F(x)) - 1]/(exp(\lambda\tau) - 1). \tag{4.5}$$

We assume here that at least one event occurs during a time interval τ. The median of this distribution μ_τ is determined as the root of

$$\Phi_\tau(x) = [exp(\lambda\tau F(x)) - 1]/(exp(\lambda\tau) - 1) = 1/2 \tag{4.6}$$

We can derive from (4.6) the median of the maximum event μ_τ for the Pareto distribution (4.1):

$$\mu_\tau = [1 - log((e^{\lambda\tau} + 1)/2)/(\lambda\tau)]^{-1/\beta}. \tag{4.7}$$

If $\lambda\tau >> 1$, i.e., the average number of events in the time interval τ is much greater than unity, then Eq. 4.7 simplifies to

$$\mu_\tau \cong (\lambda\tau/log(2))^{1/\beta}. \tag{4.8}$$

Again we remark that for $\beta < 1$ the median μ_τ increases with time in a nonlinear manner like $\tau^{1/\beta}$ with an index $1/\beta$ larger than unity.

If we do not know whether the distribution of an individual effect is the Pareto law, the sample distribution function $F_n(x)$ being only available, we can try to find a sample estimate of the median μ_τ. From Eq. 4.6 we have

$$F(x) = (\lambda\tau)^{-1} log[(e^{\lambda\tau} + 1)/2]. \tag{4.9}$$

The median of maximum loss for τ years μ_τ can be estimated by inserting into (4.9) the sample distribution function $F_n(x)$ instead of its theoretical analogue $F(x)$ with subsequent interpolation of the discrete function $F_n(x)$. We are going to determine the maximum admissible τ in such estimation. One important restriction on τ appears in this derivation: the interpolation of a sample function $F_n(x)$ assumes that its arguments do not exceed the observed maximum effect, i.e., the maximum in the catalog. This requirement is equivalent to the inequality

$$(1 + \lambda\tau) \cdot log[(e^{\lambda\tau} + 1)/2] < 1 - 1/n. \tag{4.10}$$

In other words, the catalog size n and the time interval τ of the maximum effect in question should be connected by restriction (4.10), which does not permit the median μ_τ to go "out of the data". If we accept for simplicity $n \cong \lambda T$ and $\lambda\tau >> 1$, then the restriction (4.10) can be reduced to

$$\tau < (n/\lambda) \cdot log(2) \cong 0.7T. \tag{4.11}$$

Thus, the time interval τ of a reasonable estimate of μ_τ by this method, i.e., the time interval for which the future maximum event is predicted, should not exceed 0.7 of the catalog duration T.

This restriction on τ/T can be relaxed by using a parametric approach and the limit theorems of the extreme value theory (see Chapter 6).

Now we return to the case where the distribution of an individual loss can be fitted by the Pareto law. It is enough if this is true only for observations above some threshold a. Of course, this threshold should provide a sufficiently large number of observations exceeding a, guaranteeing a reliable estimation of the Pareto index (no less than 25–30).

The probability density and the distribution function are

$$f(x) = \beta \cdot a^{\beta}/x^{1+\beta}; \quad x \geqslant a, \tag{4.12}$$

$$F(x) = 1 - (a/x)^{\beta}; \quad x \geqslant a. \tag{4.13}$$

We denote the number of observations above the threshold a as n, and the sample as $(x_1, \ldots, x_n)$. The likelihood function is

$$L = f(x_1)f(x_2)\ldots f(x_n) = \beta^n a^{n\beta}/(\Pi\, x_i)^{1+\beta}.$$

The maximum likelihood estimate (MLE) of β maximizes L:

$$\hat{\beta} = [1/n \sum log(x_i/a)]^{-1} \tag{4.14}$$

The asymptotic variance of MLE is (see, e.g., [Hi])

$$Var(\beta) = \{E(\partial logL/\partial\beta)\}^2 = \beta^2/n. \tag{4.15}$$

4.1.3 The Relation Between the Total Loss and the Maximum Individual Loss for the Pareto Law

We assume the sequence of earthquakes to be a stationary Poisson process with intensity λ events per year. The random number of events for T years is a Poisson rv with mean λT. We denote the maximum loss for this time period as M_T and the total loss as $\sum_T$. We are now going to derive some relations between M_T and $\sum_T$. It is well known (see, e.g., [Gu1]) that for distributions with a light (exponential) tail M_T increases as the logarithm of T:

$$M_T \cong c \cdot logT, \tag{4.16}$$

where c is some constant, whereas $\sum_T$ increases like T, in accordance with the CLT:

$$\sum\nolimits_T \cong T \cdot b + \zeta \cdot \sigma \cdot T^{1/2}, \tag{4.17}$$

where ζ is some standard normal rv, b is the expectation and σ the std of an individual loss. We consider the ratio $R(T)$ of total loss to maximum individual loss:

$$R(T) = \sum\nolimits_T / M_T$$

It follows from (4.16), (4.17) that

$$R(T) \cong (b/c) \cdot (T/log(T)). \tag{4.18}$$

Hence it follows that the $R(T)$ ratio increases with T linearly if we disregard the slowly varying $log(T)$. A quite different behavior of $R(T)$ occurs for heavy-tailed distributions. For such distributions $R(T)$ grows much slower and can even have a finite expectation. In other words, in this case the total and the maximum losses become comparable, i.e., the total loss is largely controlled by the maximum loss. This result justifies using, in the case of heavy tails, the *maximum* individual loss as an estimate of the total loss.

In order to illustrate this assertion, we are going to derive lower and upper bounds for $R(T)$ for the Pareto distribution of an individual loss. Taking logarithms of $R(T)$ and their expectations, we get

$$E\ log \sum\nolimits_T = E\ logR(T) + E\ logM_T. \tag{4.19}$$

Using Jensen's inequality for concave functions, we get

$$E\ logR(T) \leqslant logER(T). \tag{4.20}$$

As shown in [P4],

$$[1 - (\lambda T)^{1-1/\beta} \cdot \gamma(1/\beta; \lambda T)]/(1-\beta);\ \ \beta \neq 1; \tag{4.21}$$

$$ER(T) =$$

$$log(\lambda T) - exp(-\lambda T) \cdot (log\lambda T - 1);\quad \beta = 1, \tag{4.22}$$

where $\gamma(\alpha; t)$ is the incomplete gamma function. The function $ER(T)$ is tabulated in Table 4.1.

If $\lambda T \to \infty$, then in (4.21) the incomplete gamma function $\gamma(1/\beta; \lambda T)$ tends to the standard gamma function $\Gamma(1/\beta)$. It is possible to derive an explicit expression for $E\ log(M_T)$ [P4]:

$$E\ logM_T = [log(\lambda T) + C - Ei(-\lambda T)]/\beta, \tag{4.23}$$

Table 4.1 Expectation $E\ R(T)$ as function of λT и β

β	λT (or n)						
	10	50	100	300	500	1,000	∞
3	5.77	17.93	28.8	60.7	85.7	136.6	∞
2	4.6	11.52	16.7	29.7	38.6	55.0	∞
1.3	3.48	6.55	8.26	11.6	13.5	16.42	∞
1.1	3.1	5.16	6.15	7.85	8.7	9.91	∞
1.0	2.88	4.49	5.18	6.28	6.78	7.48	∞
0.9	2.59	3.81	4.26	4.92	5.2	5.56	10.0
0.8	2.45	3.29	3.57	3.91	4.04	4.19	4.3
0.6	2.02	2.33	2.4	2.45	2.46	2.48	2.5

where C is the Euler constant ($C = 0.577\ldots$), and $Ei(-\lambda T)$ is the integral exponential function

$$Ei(x) = \int_{-\infty}^{x} \frac{\exp(z)}{z} dz; \quad x < 0.$$

Combining equations (4.19), (4.20), (4.21), (4.22), and (4.23), we derive an upper bound on $E\ log\sum_T$:

$$E\ log\sum\nolimits_T \leqslant 1/(1-\beta)\cdot[1-(\lambda T)^{1-1/\beta}\cdot\gamma(1/\beta;\lambda T)] + [log(\lambda T) + c - Ei(-\lambda T)]/\beta. \tag{4.24}$$

If $\lambda T >> 1$, then (4.24) can be simplified:

$$Elog\sum\nolimits_T \leqslant \begin{cases} [1-(\lambda T)^{1-1/\beta}\cdot\Gamma(1/\beta)]/(1/\beta) + [log(\lambda T) + C]/\beta; \quad \beta \neq 1, & (4.25) \\ log(\lambda T) + log(log\lambda T) + C; \quad \beta = 1. & (4.26) \end{cases}$$

From Eq. 4.21 it is seen that if $\beta < 1$ and $\lambda T >>1$, then

$$ER(T) \cong (1-\beta)^{-1}, \tag{4.27}$$

and if $\beta > 1$ and $\lambda T >>1$, then

$$ER(T) \cong (\beta - 1)^{-1}\cdot(\lambda T)^{1-1/\beta}\Gamma(1/\beta). \tag{4.28}$$

Thus, if $\beta < 1$, then the random quantities $log\Sigma_T$ and $log\mathrm{M}_T$ are comparable in value and both grow as $log(\lambda T)^{1/\beta}$. This fact might be interpreted as a nonlinear growth of Σ_T and M_T at the same rate as for $(\lambda T)^{1/\beta}$. We recall that both random quantities have infinite expectations, if $\beta < 1$.

If $\beta > 1$ and $\lambda T >> 1$, then $log[ER(T)]$ and $log M_T$ increase with T as follows:

$$log[ER(T)] \sim (1 - 1/\beta) \cdot log(\lambda T), \tag{4.29}$$

$$E\ log\ M_T \sim (1/\beta) \cdot log(\lambda T). \tag{4.30}$$

It can be seen that, if $\beta >> 1$, then $ER(T)$ increases "almost linearly", whereas M_T increases very slowly, like $T^{1/\beta}$.

In order to derive a lower bound on the total loss, we rewrite $R(T)$ in the following form:

$$R(T) = \sum\nolimits_T / M_T = \sum_{i=1}^{n} x_i / M_{T,} \tag{4.31}$$

where n is a random Poisson variable with parameter λT, and x_i are the losses. Since the geometric mean never exceeds the arithmetic mean, we have from (4.31):

$$\begin{aligned} logR(T) &= logn + log\left(1/n \cdot \sum_{i=1}^{n} x_i/M_T\right) \geqslant logn + log\left(\prod_{i=1}^{n} x_i/M_T\right)^{1/n} \\ &= logn + 1/n \sum_{i=1}^{n} logx_i - logM_T \end{aligned} \tag{4.32}$$

Taking expectation of both sides of (4.32) and integrating the second term, we get

$$E\ logR(T) \geqslant E\ logn + 1/\beta - E\ logM_T. \tag{4.33}$$

Inserting (4.33) into (4.20), we get a lower bound on $E\ log \sum_T$:

$$E\ log \sum\nolimits_T \geqslant E\ logn + 1/\beta. \tag{4.34}$$

For the Poisson sequence of events we have

$$E\ logn = \sum_{k=1}^{\infty} (logk) \cdot exp(-\lambda T) \cdot (\lambda T)^k / k! \tag{4.35}$$

Using the fact that n is asymptotically normal with mean λT and std $(\lambda T)^{1/2}$, we get

$$E\ logn \cong log\lambda T. \tag{4.36}$$

Inserting this expression into (4.34) we get a lower bound on $E\ log\sum_T$ for the case
$\lambda T >> 1$:

$$E\ log\sum\nolimits_T \geqslant log(\lambda T) + 1/\beta. \tag{4.37}$$

The approximation (4.36) is applicable only if $\beta < 1$. If $\beta > 1$, then $E\ log\sum_T$ has $E\ log\mathrm{M}_T$ for its lower bound:

$$E\ log\sum\nolimits_T \geqslant E\ logM_T = \{log(\lambda T) + C - Ei(-\lambda T)\}/\beta. \tag{4.38}$$

Thus, the lower bound on $E\ log\sum_T$ is given by (4.34), (4.37) for $\beta > 1$, and by (4.38) for $\beta < 1$. If the parameter β is estimated from a finite sample, and is subject to random scatter, we can evaluate the resulting uncertainty by putting the lower and upper bounds of β into the equations that give the bounds of $E\ log\sum_T$.

4.2 The Growth of Total Earthquake Loss

Now consider the behavior of total earthquake loss as seen in actual observations. These observations can be considered as satisfactorily complete for the period 1900–1999 (casualties) and 1960–1990 (economic losses), although there are some fragmentary data for earlier periods.

As we have shown earlier, the earthquake loss distribution can be quantified as a distribution with a heavy tail, which should lead to a nonlinear growth of total loss. Perhaps another possible cause of this nonlinearity could be the nonstationarity of the time series of losses caused by the evolution of socio-economic structure or the nonstationarity of the seismic process. But we are going to show that these factors are secondary in importance. The main cause of the nonlinear growth of losses can be explained by the heavy-tailed character of loss distribution. We provide a tentative forecast of the world earthquake losses for the future 50 years, assuming that all statistical characteristics of the process will retain the same behavior as in 1900–1999. The forecast incorporates both the existence of different modes of growth for the total loss and the dependence of this mode on socio-economic conditions.

4.2.1 The Raw Data on Seismic Disasters

We used for our analysis the most complete data on the worldwide earthquake losses as provided by the US National Data Center (http://www.neic.cr.usgs..gov/neis/eqlists, accessed in 2007). This document is a compilation of various data on earthquake casualties and losses taken from many sources, in particular [Lo, NG, KS3] and others.

For some events there are loss data from several sources (which are frequently rather divergent). We shall use the data for the time period 1900–1999 as being the most reliable. When there were discrepancies in the data, we took the largest figures. We believe that this approach provides more reliable information. We used some other data sources as controls, and verified that our final conclusions depend little on the particular data set used.

The advantage of this approach to the compilation of our working data set consists in its greater completeness and in the possibility of using data for later time periods, whereas one drawback to this approach is possible inclusion of inaccurate figures. As compared with more carefully selected data in [NG] covering the period 1900–1979, our data set is much more complete, but its "contamination" is somewhat higher. We used the data of [NG] to check the inferences based on our data set.

The working catalog thus made contained 1,137 events with nonempty casualty data and 194 events with nonempty economic loss data. The total casualty figure for this period is about 3 millions. Individual, casualty data vary from 1 to as much as 650,000 (an expert estimate of the death toll in the Tang-Shan, China earthquake of July 27, 1976, $M = 7.8$). This earthquake is an impressive example of possible data discrepancies: according to the official report, there were 240,000 dead (the official figure, as will be seen below, is more consistent with the relationship of casualties and losses to socio-economic conditions).

For some events we give in the catalog merely information of the type "several dead". The contribution of such events into the total sum is small, and we have neglected these cases. The earlier and weak earthquakes provided less accurate data. We shall see that the inaccuracy does not influence significantly our final conclusions.

The economic loss data are generally less complete than the casualties. They usually refer to the time period after 1950. The discrepancies in the economic loss data are higher than for casualties. There are 194 events in our catalog for which economic losses have been reported, with the total loss amounting to more than US $130,000,000. More detailed information on earthquake losses can be found in Table 4.2 (see also Fig. 4.1).

We have taken into account the variation in the rate of exchange of the US dollars with respect to the 1990 prices and introduced a corresponding correction. The correction makes for a greater homogeneity of the catalog, but does not affect significantly the character of data. In comparable prices the worldwide economic loss amounts, for the entire period 1900–1999, to about US$200,000,000. In Table 4.2 the fraction of discrepant data are shown as characteristics of completeness and accuracy.

Table 4.2 Characteristics of the working catalog of casualties and losses together with divergences of different sources

	North America 180–60°W, 30–70°N	Europe 20°W–30°E, 35–60°N	Japan 125–160°E, 25–45°N	Latin America 45–120°W, 30°N– 60°S	Asia 30–125°E, 20–50°N	Indo-China 95°E–160°W, 20°N–20°S
Totlal number of events	61	161	75	239	376	128
Number of events with casualty data	33	155	74	236	375	125
Percentage of events with different casualty numbers	40	22	18	14	17	10
Median of ratio max/min casualties 1900–1950	3	20	2	10	5	10
The same for post-1950	1.1	1.2	1.1	4	2	3
Number of events with casualty data[a]	55	26	5	38	18	14
Percentage of events with different loss estimates	28	27	40	37	30	7
Median of ratio max/min loss 1900–1950	8	7	5	20	100	–
The same for post-1950	1.3	2	–	2	2	3

[a]Without events with merely qualitative gradations of death toll: several, small, moderate, etc.

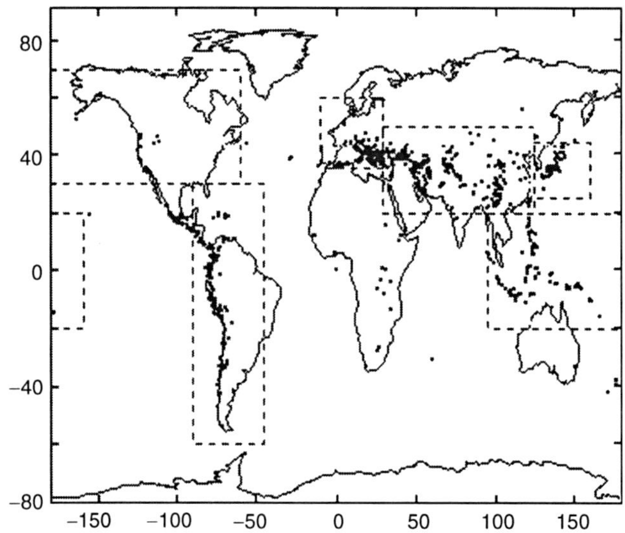

Fig. 4.1 Epicenter locations for events with known loss (*dots*). Region boundaries are shown by *dotted lines* (Europe, South America, etc.)

4.2.2 The Nature of Nonlinear Growth of Cumulative Earthquake Loss

As noted earlier, the cumulative, total loss from earthquakes exhibits a tendency to a nonlinear growth. This is also the case for our data. Figure 4.2a displays, in a log–log plot, the time series of cumulative casualties and the time series of cumulative numbers of earthquakes with non-zero victims. We fitted these time series with the function t^{α}, getting $\alpha = 1.7$ for cumulative casualties and $\alpha = 1.3$ for the number of earthquakes. We see that these time series are increasing over time in a nonlinear manner, the former increasing on an average appreciably faster than the latter. It also appears from Fig. 4.2a that the plot of the cumulative number of casualties has a lower slope by the end of the observation period; the causes of this effect will be discussed later.

Similar patterns are obtained for the Nelson–Ganze catalog, but the exponent α is slightly smaller; this may be explained by the procedure employed to compile the working catalog, where the maximum numbers of casualties and maximum losses have been included.

Figure 4.2b shows similar data on earthquake losses. The catalog becomes more complete as time goes on, the completeness requirement making any data analysis possible for the post-1950 period only. Nevertheless, the same patterns can be seen in this short time span as those for the number of casualties. Loss values are increasing non-uniformly and generally faster than the number of the responsible earthquakes. In the 1960–1970s the rate of increase for the cumulative loss reaches

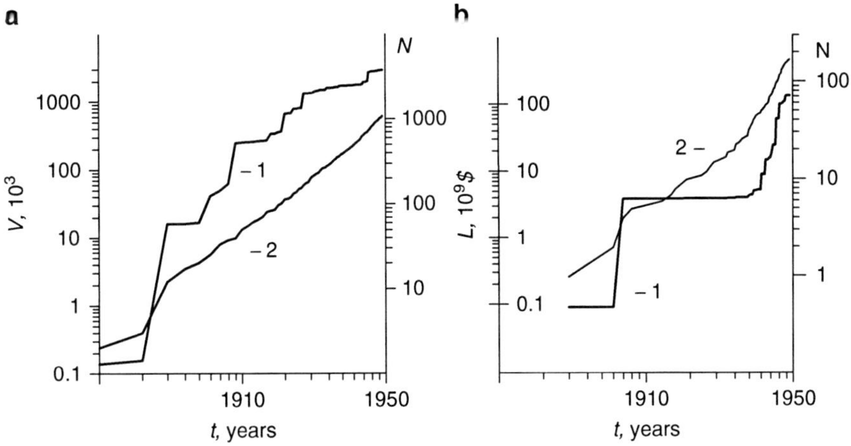

Fig. 4.2 Nonlinear time-dependent growth of casualties V, in thousands (**a**), and economic losses L, billion dollars (**b**). 1 – Number of casualties (a); loss (b); 2 – Number of events with non-zero casualties or loss

the maximum, the exponent α approaching 6(!). The increase of cumulative loss was decelerated toward the end of the time span under analysis. Of course, the above results are preliminary, in particular, because the estimate of the exponent is strongly dependent on the chosen time origin. There is a variety of factors to cause non-linearities in the growth of cumulative numbers of casualties and economic loss. On the one hand, the non-linearity may result from a progressive growth of earthquake losses over time. This kind of non-stationarity seems quite possible as a result of increasing population and expanding infrastructure, as well as of the abovementioned increase in the rate of damaging earthquakes. On the other hand, a nonlinear growth of the cumulative loss can also occur in a stationary model, specifically, when the distribution has a heavy tail.

We shall analyze stationarity in the loss data by examining the earthquakes causing casualties in three different ranges:

Range I: 1–9 casualties, a total of 412 events
Range II: 10–99 casualties, a total of 359 events
Range III: 100 or more casualties, a total of 366 events

It appears that Range III includes a greater part of the casualties. As a matter of fact, the events in Range III are more important, since the few extreme disasters make a very high contribution. Overall, 0.7% of all casualties were due to Range II, the figure for Range I being lower still (0.07%).

Figure 4.3 shows how the cumulative numbers of events in the different ranges increase with time. One can see that the sequence of Range III events is practically stationary in the entire time span. The rate of events in Ranges I and II is increasing in time, especially Range I. Toward the end of the time span of interest the events in Ranges I and II became more stationary.

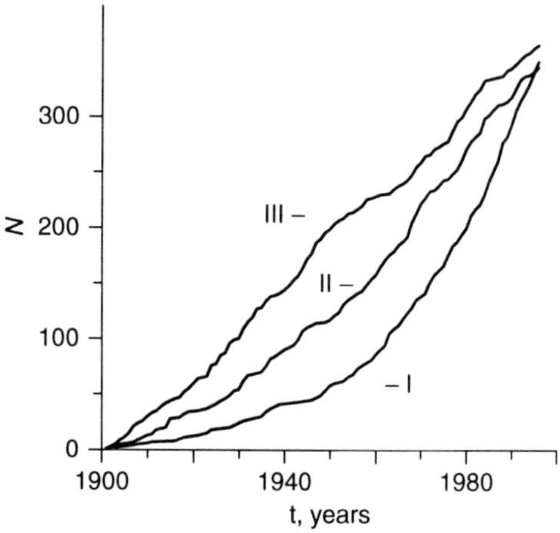

Fig. 4.3 Growth of cumulative number of earthquakes for three ranges of casualties: I – 1–9 casualties; II – 10–99; III – more than 99

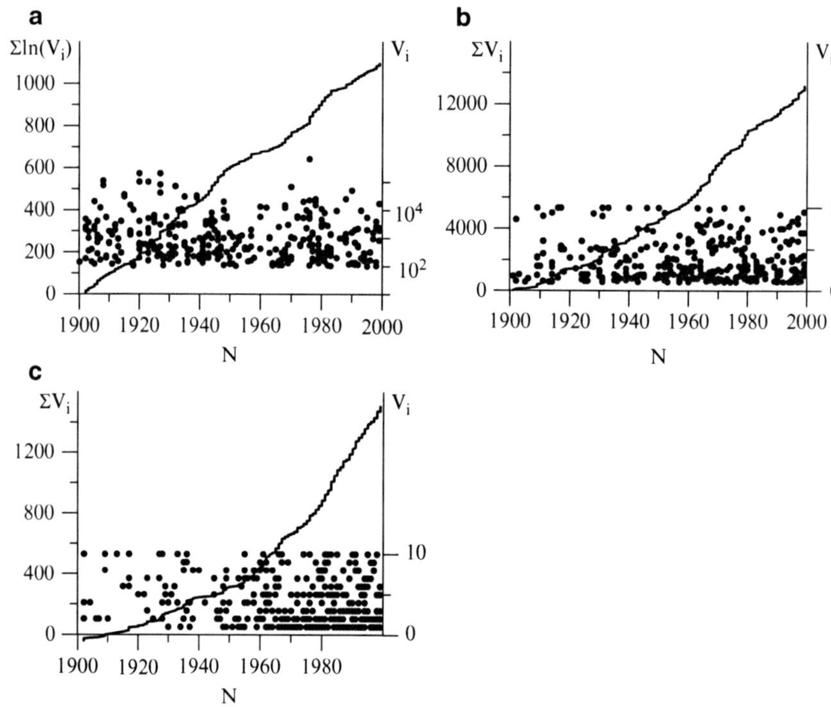

Fig. 4.4 Time series of the number of casualties (*dots*) and cumulative sums (*line*) for three ranges of casualties. (**a**) Events in range I; (**b**) events in range II; (**c**) events in range III, in case (**a**) cumulative sums are given on a log scale

We tried to find out how the distribution of the number of casualties varies over time in different ranges by plotting both cumulative and individual casualties versus time for all the three ranges, see Fig. 4.4. For Range III we took logarithms, since the corresponding casualties differ by three orders and more, so it is more convenient to check stationarity using log casualties. For a stationary time series the plot of its cumulative sum should not deviate considerably from a straight line. From Fig. 4.4a it can be seen that the cumulative sum of the logarithms of casualties for Range III shows no significant deviations from a straight line, indicating that the mean logarithm of the number of casualties is constant. Hence one can conclude that the distribution of the number of casualties for Range III is stationary. The cumulative casualties of Range II, Fig. 4.4b, exhibit somewhat faster growth toward the end of the time span. The distribution of casualties of Range I events, Fig. 4.4c, does vary in time and tends to decrease as time goes on, whereas the mean number of victims per event has a tendency to decrease. Such changes can be explained by increasing improvements in the recording network. We see that non-stationarities in the behavior of seismic disasters mostly occur for smaller events (Range I), and perhaps, to a lesser extent, for moderate events (Range II), which contribute less than 1% in the total casualties. Thus, a significant nonlinear growth of cumulative casualties stems from other causes than non-stationarities in the sequence of seismic disasters.

It is reasonable to conclude the above discussion of non-stationarities in the occurrence of earthquake losses by asking what is the relation, if any, between losses and seismicity rate variations. We have compared time variations in the rate of large earthquakes and associated energy release in relation to the number of casualties. The series of the numbers of casualties and earthquake energy release per unit time were subject to very large variations. Bearing this in mind, we have compared strongly smoothed values of the number of casualties and energy release. As was to be expected, the correlation between these two time series was not significant (the correlation coefficient was $r = 0.44$).

We now summarize our analysis of variability for the occurrence of earthquake disasters. The disasters in Range III, which make the dominant contribution into total earthquake loss, occur in a quasi-stationary manner. The disaster occurrence is non-stationary for comparatively smaller events in Ranges I and II, a total of less than 1% of all casualties, based on worldwide data. The growth in the rate of smaller disasters identified here can most naturally be explained by improved detection and reporting levels; this point of view is corroborated by the fact, to be noted below, that these events began to occur in a quasi-stationary manner earlier in developed countries than elsewhere. The fundamental fact of a strong nonlinear growth of the cumulative earthquake losses can thus be explained neither by seismicity rate variations nor by a non-stationary occurrence of earthquake disasters.

We are going to show that the nonlinear growth of cumulative casualties can be explained on the basis of a stationary model, when account is taken of a heavy tail in the distribution of earthquake losses. In accordance with the conclusion drawn above, the distribution of casualties exceeding some sufficiently high threshold a can be fitted to good accuracy by a power-law distribution, the Pareto law (4.13). The MLE of the parameter β is

$$\hat{\beta} = \{1/n \sum log(x_i/a)\}^{-1}. \tag{4.39}$$

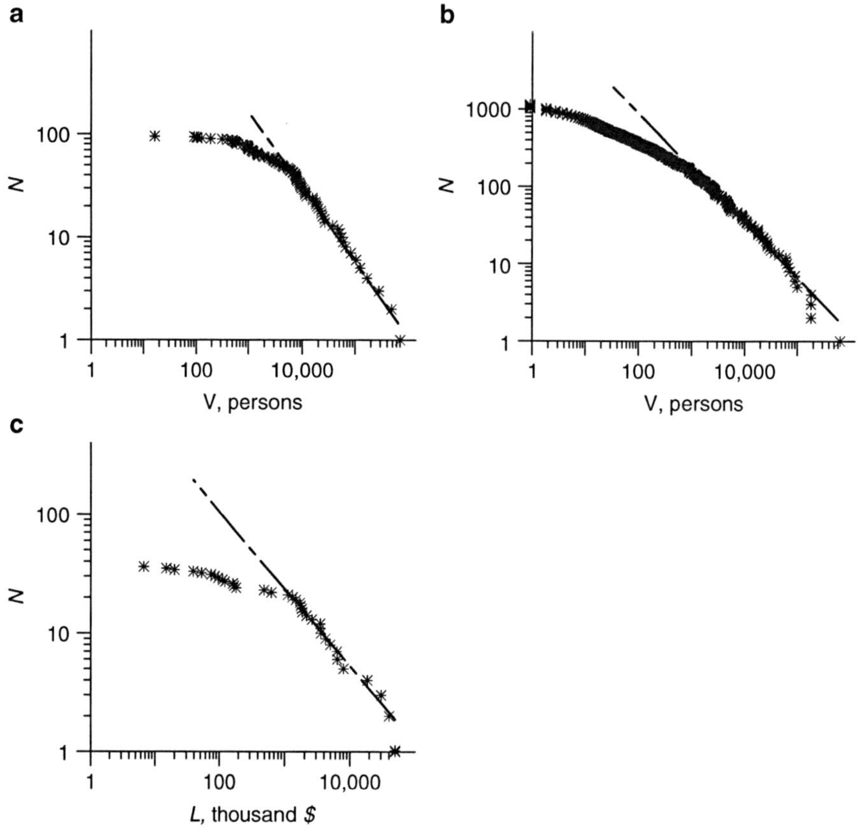

Fig. 4.5 Non-normalized complementary sample distribution function for: (**a**) annual numbers of casualties; (**b**) casualties for individual event; (**c**) annual economic losses. The fitted power-law functions are shown by *dotted lines*, with $\beta = 0.77 \pm 0.11$ (**a**); $\beta = 0.73 \pm 0.11$ (**b**); $\beta = 0.65 \pm 0.16$ (**c**)

The std of the estimate equals

$$\mathrm{Std}(\hat{\beta}) = \beta / n^{1/2}. \tag{4.40}$$

The complementary distributions (tail distributions) of the earthquake casualties and losses are presented in Fig. 4.5: annual casualties (a), casualties for individual events (b), annual losses (c). The distribution of casualties for 150 greatest disasters (96% of total casualties) is satisfactorily described by the Pareto law with $\beta = 0.71 \pm 0.06$. If we assume the times t_i of the disasters to be Poissonian, we get the so-called Cramer–Lundberg model [EKM] where the cumulative effect is given by the sum

$$\sum\nolimits_t = \sum_{t_i \leqslant t} x(t_i) \tag{4.41}$$

For the 47 largest annual casualty sums (97% of total casualties) the Pareto parameter is $\beta = 0.76 \pm 0.11$. In both cases $\beta < 1$, in other words, both the event-approach and the year-approach lead to a heavy-tailed distribution. Both models yield similar results. The event-approach has a larger sample size, and somewhat smaller β. We shall use both approaches for control purposes.

Figure 4.5c shows similar distributions for annual economic losses. We used here the post-1960 data, which can be considered as more or less stationary. For annual losses we have $\beta = 0.65 \pm 0.16$, and the Pareto law fits well the empirical distribution for 21 largest losses of 40. Taking into account the small sample size (1960–1999), we estimated β separately for the post-1970 data. We got $\beta = 0.70 \pm 0.17$. The thresholds were the same for both cases: $a =$ US$1,200,000,000. The fitting of an empirical distribution by the Pareto law can be considered satisfactory when working above this threshold.

We have thus fitted, for all options considered, the casualty distribution and the loss distribution by the Pareto law with $\beta < 1$. We can expect that the time-dependent growth of the losses will exhibit nonlinear features discussed above. In particular, the total effect can to a large extent be determined by an individual event, viz., the maximum. The sample means are unstable and non-informative.

Here, however, we should repeat that in real situations there are always some restrictions of a general, global character. Evidently, the number of victims cannot exceed the Earth's population, neither the economic loss can exceed the total cost of the whole thechnoshpere. Thus, the Pareto law cannot represent the actual process for arbitrarily long time spans. The distribution of any actual loss should be truncated, or at least begin to decrease very rapidly, at some very high threshold A. If the process can be observed over a very long time span, when events of size A (or close to A) begin to appear repeatedly, then the character of growth of cumulative losses will change: the Law of Large Numbers begins to be applicable, and nonlinear effects gradually disappear. Thus, it is very important to be able to estimate the threshold A and the time span of "validity of non-linear regime". Some approaches to this problem are suggested below.

Sometimes, treatment of heavy-tailed data can be facilitated by using logarithms of original values, as we did earlier. The passage to logarithms, which can be done when the original numerical values are positive, ensures that all theoretical moments exist, hence the Law of Large Numbers and the Central Limit Theorem for sums of logarithms are applicable. Occasionally however, in particular, for evaluation of total loss, interest focuses on the sum of original values themselves rather than the sum of their logarithms. The latter cannot be used, because there will be no backward passage from a sum of logarithms to that of original values. In that case some other limit theorems of the theory of probability can be used. We know that properly normalized and centered sums of heavy-tailed variables can converge to a so-called stable law. A stable law is characterized by a positive index α, $0 < \alpha \leq 2$. If $\alpha = 2$, then the limit distribution is normal. If $1 < \alpha < 2$, then the expectation exists (and so the Law of Large Numbers is applicable), but the variance and std are infinite. If $0 < \alpha \leq 1$, then the expectation is infinite and the limit distribution has a power-like tail equivalent to $1/x^{\alpha}$. In order to characterize a typical middle value for such distributions the median is used.

We will now derive theoretical estimates of cumulative casualties for N years, $\sum_{N,}$ when the individual loss obeys the Pareto law with index $\beta < 1$. Since we use only observations X above a fixed threshold a, some years will be empty, i.e., the number of years with nonempty casualties n will be a binomial random variable with the parameter p:

$$p = Pr\{X \geqslant a\}; \quad Pr\{n = j\} = C_N^j p^j (1-p)^{N-j}; \quad j = 0, 1, \ldots, N. \tag{4.42}$$

We begin by assuming that n is non-random. The normalized sum $\sum_n/n^{1/\beta}$ converges to a random variable having a stable distribution with index $\alpha = 1/\beta$. We emphasize that the sum $\sum_n$ is normalized by $n^{1/\beta}$ instead of the ordinary $n^{1/2}$ used in the CLT. We denote the maximum value in the sample as M_n, $M_n = max(x_1, \ldots, x_n)$, and define the ratio R_n

$$R_n = \sum\nolimits_n / M_n.$$

The R_n ratio is tabulated in Table 4.1, it is a random variable of order one. As a rough approximation, we can replace R_n with its expectation (see [Fe]) in (4.42), which tends to $1/(1-\beta)$ as $n \to \infty$. We get

$$\sum\nolimits_n \cong \mathrm{M}_n/(1-\beta). \tag{4.43}$$

The median Med_n of M_n can be found explicitly, using the fact that the maximum value in a sample of size n has the distribution function $F^n(x)$. We find

$$Med_n = a \cdot [1 - 2^{-1/n}]^{-1/\beta} \cong n^{1/\beta} \cdot a \cdot (log2)^{-1/\beta}. \tag{4.44}$$

which is quite similar to Eq. 4.14. Then, substituting Med_n for M_n into (4.43), we get from (4.43) an estimate $Med\sum_n$ of the median of $\sum_n$;

$$Med\sum\nolimits_n \cong n^{1/\beta} \cdot (a/(1-\beta)) \cdot (log2)^{-1/\beta}.$$

This is an explicit approximate estimate of the median of cumulative losses. However, as has been remarked above, any point estimate of cumulative effects with heavy tails is partially depreciated by a high scatter of $\sum_n$. A more appropriate characteristic of $\sum_n$ is provided by a confidence interval, i.e., an interval $(z_\varepsilon; Z_\varepsilon)$ that encloses the normalized random variable $\sum_n/n^{1/\beta}$ with a given probability $(1-2\varepsilon)$:

$$Pr\{z_\varepsilon < \sum\nolimits_n / n^{1/\beta} < Z_\varepsilon\} = 1 - 2\varepsilon. \tag{4.45}$$

How is one to find the endpoints of the confidence interval (z_ε; Z_ε)? For large z we use the asymptotic relation (see [Fe])

$$Pr\{z < \sum_n / n^{1/\beta}\} \cong (a/z)^{1/\beta}. \tag{4.46}$$

Equating the last expression to ε, we get an approximate upper bound Z_ε:

$$Z_\varepsilon = a \cdot \varepsilon^{1/\beta}. \tag{4.47}$$

The lower bound is derived as follows:

$$Pr\left\{\Sigma_n \big/ n^{1/\beta} < z\right\} \leqslant Pr\left\{\max_{i \leqslant n}(x_i)/n^{1/\beta} < z\right\} = Pr\left\{\max_{i \leqslant n}(x_i) < n^{1/\beta} z\right\}$$
$$= [1 - (a/zn^{1/\beta})^\beta]^n \approx exp\{-(a/z)^\beta\}.$$

Equating the last expression to ε, we get an approximate lower bound z_ε:

$$z_\varepsilon = a[log(1/\varepsilon)]^{-1/\beta}, \tag{4.48}$$

and so we have derived the approximate confidence interval (z_ε; Z_ε) with confidence probability at least (1 – 2ε).

Now we take into account the fact that the number n of events above the threshold a is a binomial random quantity. Denoting the random loss for N years as L_N and using Eq. 4.42, we get

$$Pr\{L_N < x\} = \sum_{n=0}^{N} Pr\{\Sigma_n < x\} C_N^j p^j (1-p)^{N-j}. \tag{4.49}$$

Now we approximate again $\sum_n$ by $M_n/(1 - \beta)$, which has the distribution function $F^n(x{\cdot}(1 - \beta))$. We thus get the approximate equality

$$Pr\{\Sigma_N < x\} \cong F^n(x \cdot (1-\beta)) = [1 - a^\beta \cdot (x \cdot (1-\beta))^{-\beta}]^n.$$

Inserting this expression into (4.49), we get

$$Pr\{L_N < x\} \cong [1 - p + p \cdot F(x \cdot (1-\beta))]^N = [1 + p \cdot a^\beta \cdot (x \cdot (1-\beta))^{-\beta}]^N. \tag{4.50}$$

Now we can obtain the median $MedL_N$ of cumulative loss L_N for N years by equating the rhs in (4.50) to 1/2 and solving this equation:

$$MedL_N \cong ap^{1/\beta}/[(1-\beta)(1-2^{-1/N})^{1/\beta}]^{-1} \cong a(Np)^{1/\beta}/[(1-\beta)(log2)^{1/\beta}]. \tag{4.51}$$

It follows from (4.51) that, when n is random, we can use Eq. 4.44, derived under the assumption of a fixed n, and replace n with its mean $p{\cdot}N$.

The median and the 80% confidence interval for cumulative casualties $\sum_n$ based on (6.44)–(6.49) with $\beta = 0.77$ are shown in Fig. 4.6a. The theoretical median is in good agreement with the plot showing the cumulative number of casualties during a time interval of about 40 years (until 1940), but appears to be an overestimate for later years. The latter circumstance is related, as will be shown below, to the fact that the total loss must be bounded. Figure 4.6b shows similar curves for cumulative losses. It is easy to see that the respective results are qualitatively similar, but in this case the theoretical cumulative losses begin to exceed observed losses earlier, in about 15 years.

These results suggest that the demonstrated growth of cumulative earthquake losses can be explained (at least the bulk of it) in the framework of a stationary model, taking into account the observed distribution of casualties and economic losses both in the event form and in the annual form. These distributions possess heavy tails, which produces a nonlinear growth of the type $t^{1/\beta}$, $\beta < 1$.

A full solution to the problem of describing this behavior should include, however, two more questions. In the first place, it is necessary to establish the limits to an admissible fitting of the loss distribution by the Pareto law, and, accordingly, the time span of nonlinear growth. In the second place, it is desirable to clarify the character of interrelations between earthquake losses and socio-economical processes, such as population growth and the development of the infrastructure. The former question will be considered in the next chapter, the latter in Chapter 7.

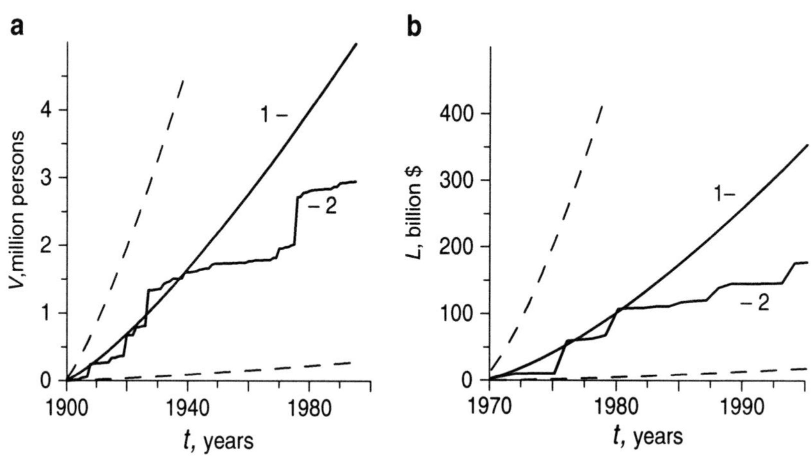

Fig. 4.6 Actual and theoretical curves of the cumulative growth of casualties (**a**) and economic losses (**b**). 1 – Median of casualties (a) or economic losses (b); *dotted lines* are the upper and lower 80% confidence bounds; 2 – actually recorded data: the number of casualties (since 1900) or economic losses (since 1970)

4.2.3 The Limits of Applicability of the Pareto Law to the Estimation of Earthquake Losses

That the actual losses are bounded means that there must be restrictions on the use of the Pareto power law (4.1) with $\beta < 1$. A simple model with an upper bound is provided by the truncated Pareto law with the distribution function $F(x)$:

$$F(x) = [1 - (a/x)^{\beta}]/[1 - (a/A)^{\beta}], \quad a \leqslant x \leqslant A. \tag{4.52}$$

One could replace this sharp cutoff by some gradual versions that make the density $f(x)$ decay rapidly enough at infinity. To take an instance, Kagan [K2] proposed using the gamma distribution to fit the scalar seismic moment data. He introduced into the power-law tail an exponential term that ensures the existence of all moments. Sornette et al. [SKKV] suggested using a second branch of the Pareto law, with $\beta > 1$, which ensured finiteness for the mathematical expectation of losses. These authors pointed out that the crossover point x_c can usually be estimated from experimental data with considerable uncertainty. We have dwelt on the model (4.52) because this affords a better review of the effects related to the limitedness. We shall assume the upper boundary A to be much higher than the threshold a, i.e.,

$$A/a >> 1.$$

It can be expected that the truncation of the Pareto distribution as defined in (4.52) will begin to affect the statistical behavior of the cumulative sum $\sum_n$ only for large enough n. When n is small, $\sum_n$ is not much affected by the truncation of the Pareto law. A rough estimate of the critical n can be found as follows. Consider the probability of the following event: all n values in a sample drawn from the unlimited Pareto law are less than A:

$$Pr\{x_1 < A, \ldots, x_n < A\} = [1 - (a/A)^{\beta}]^n. \tag{4.53}$$

This probability tends to zero as $n \to \infty$, but for small n it is still large. Demanding this probability to be larger than $(1 - \varepsilon)$ with a small ε, we get a restriction on n:

$$[1 - (a/A)^{\beta}]^n > 1 - \varepsilon,$$

$$n\log[1 - (a/A)^{\beta}] > \log(1 - \varepsilon),$$

$$-n(a/A)^{\beta} > \log(1 - \varepsilon),$$

$$n < (A/a)^{\beta}\log(1/(1 - \varepsilon)). \tag{4.54}$$

The conditional distribution of $(x_1, \ldots, x_n)$ given $max(x_1, \ldots, x_n) < A$ for the unlimited Pareto coincides with the distribution for the truncated Pareto, and the probability of this restriction is high, $(1 - \varepsilon)$. Thus, with probability larger than $(1 - \varepsilon)$, the behavior of $\sum_n$ will be the same both for the unlimited and for the truncated Pareto.

We now are going to derive a somewhat more accurate estimate of the restriction on n based on the median Med_n of maximum M_n. We solve the equation

$$[(1 - (a/x)^{\beta})/(1 - (a/A)^{\beta})]^n = 0.5,$$

and find the median Med_n:

$$Med_n = a/\{1 - 2^{-1/n}[1 - (a/A)^{\beta}]\}^{1/\beta}. \tag{4.55}$$

Comparing (4.55) with the median for unlimited Pareto (4.44), we see that the difference between the two increases with n increasing. We can take as the critical n the value that makes the ratio of the two medians (4.55)/(4.44) no less than 0.9. Then we get for n the restriction

$$n \leqslant N_1 = 1/log_2(1 + [1 - (1 - r)^{\beta}]/[0.9^{-\beta} - 1]), \quad r = a/A. \tag{4.56}$$

On the contrary, if $n >> N_1$, then for the cumulative sum of the truncated Pareto rv one can use CLT and the Gaussian distribution

$$(\Sigma_n - nEx_1)/\sqrt{nVar(x_1)} \cong \eta, \tag{4.57}$$

Where η is a standard Gaussian variable and:

$$Ex_1 \cong A(a/A)^{\beta}\beta/(1 - \beta);$$

$$Var(x_1) \cong A^2(a/A)^{\beta}\beta/(2 - \beta). \tag{4.58}$$

As to the critical upper bound for n, we can proceed as follows. For truncated Pareto the ratio $\rho = E\,\Sigma_n /(Var(x_1))^{1/2}$ tends to infinity as $n \to \infty$, whereas for the unlimited Pareto it tends to zero. It follows that we can take as the critical n some threshold value ρ_c for the ratio ρ: $\rho \leq \rho_c$.We take $\rho_c = 3$. Thus, we obtain the following restriction on n:

$$n \geqslant N_2 = 9(1 - \beta)^2(A/a)^{\beta}/[\beta/(2 - \beta)]. \tag{4.59}$$

We point out here that N_2 has the same leading term $(A/a)^{\beta}$ as that in (4.56). Practical applications of (4.56), (4.59) require estimates of the parameters β and A. Consider the case of annual values of the casualties. It can be seen from Fig. 4.5a that the annual rate of casualties is satisfactorily described by the truncated Pareto distribution (4.52) for $x > 6{,}000$. So we take $a = 6{,}000$. This threshold has been exceeded 47 times during 100 years, which means that values $x > 6{,}000$ appear with the frequency $p \cong 0.47$ in the series of annual observations. The parameter β can be replaced with the maximum likelihood estimate (4.40). As to the parameter A, any methods used to estimate it are likely to involve large uncertainties, because the largest events are rare. Based on the largest annual number of casualties in the catalog, 680,000 victims, one can put $A = 1{,}000{,}000$. Then we get from (4.56), (4.59):

$$N_1 = 13; \quad N_2 = 26. \tag{4.60}$$

Now it should be taken into account that the losses we have used are observed only in 46% of all years (when the number of victims is larger than 6,000). Hence the numbers $N_1 = 13, N_2 = 26$ should be approximately doubled. We conclude that a nonlinear growth of cumulative casualties can be detected for time spans up to 20–25 years. The sample means and std used in standard data processing is applicable only to time spans larger that 50 years. A transitional mode of loss growth should be expected in the interval from 25 to 50 years.

The largest uncertainty in the estimation of N_1, N_2 is caused by the choice of A. One can be certain that at the present time there is no satisfactory method for evaluating the maximum possible amount, either of earthquake losses or of earthquake energy. All existing methods are quite uncertain and sometimes semi-qualitative, basically because of insufficient time spans of the catalogs, whereas the estimation of the maximum possible value of a random variable requires a very long series of observations which is not generally the case. Still, because of the importance of A, we dwell on this problem in more detail. We shall suggest two simple estimators of A, N_1, N_2, and check how these important parameters vary from catalog to catalog. Other approaches to this problem will be discussed in Chapters 5 and 6.

We shall make use of the Cramer–Lundberg model [EKM], which uses available information more fully than the aggregated annual loss model does.

The first estimator is based on the so-called fiducial approach (see [P2, PL]). The conditional probability density of loss X given $X < a$ is denoted $g(x)$. The mean of X given $X < a$ is small compared with the expectation of X. Therefore, the particular form of $g(x)$ to be used is immaterial. We assume that the density and distribution function of losses are

$$\varphi(x;A) = \begin{cases} (1-p)g(x); & x < a \\ p\beta a^{\beta}/\{x^{(1+\beta)}[1+(a/A)^{\beta}]\}; & a \leqslant x \leqslant A \end{cases} \tag{4.61}$$

$$\Phi(z,A) = \int_0^z \varphi(x,A)dx = 1 - p + p[1-(a/z)^{\beta}/[1-(a/A)^{\beta}], \quad z > a. \tag{4.62}$$

The maximum of a sample $(x_1, \ldots, x_n)$, denoted as $m_n = max(x_1, \ldots, x_n)$, has the distribution function $\Phi^n(z, A)$:

$$Pr\{m_n < z\} = \Phi^n(z, A). \tag{4.63}$$

Now we set forth the fiducial approach in formal terms (for details see [P2]). We denote by $U_A(x)$ the inverse function with respect to $\Phi(z, A)$ with a fixed A:

$$U_A(\Phi(z, A)) \equiv z;$$

$$\Phi(U_A(z), A) \equiv z.$$

We denote further:

$$t = \Phi(z, A).$$

Then $z = U_A(t)$, and we have from (4.63)

$$Pr\{m_n < U_A(t)\} = t^n. \tag{4.64}$$

Now we introduce a function $V_t(A)$ which is the inverse of $U_A(t)$ under fixed t:

$$V_t(U_A(t)) \equiv A;$$

$$U_A(V_t(A)) \equiv A.$$

Applying the function $V_t(A)$ to both sides of the inequality $m_n < U_A(t)$ in (4.64), we get

$$Pr\{V_t(m_n) < A\} = [\Phi(z, A)]^n. \tag{4.65}$$

Setting $z = m_n$, $t = \Phi(m_n, A)$ in (4.65), we get

$$Pr\{V_t(m_n) < A\} = [\Phi(z, A)]^n. \tag{4.66}$$

Finally, taking the complementary probability to (4.66), we get formally:

$$Pr\{A \leqslant V_t(m_n)\} = 1 - [\Phi(m_n, A)]^n, \tag{4.67}$$

The probability (4.67) can be considered (at least formally) as the distribution function of A, although A is not necessarily a random variable. The distribution function (4.67) should be completed by an ideal point $A = \infty$ with the probability

$$Pr\{A = \infty\} = [\Phi(z, A)]^n. \tag{4.68}$$

The distribution (4.67), (4.68) is strongly skewed. Equation 4.68 implies an infinite expectation. That is why we estimate the "typical value" of A by the median $M_n(A)$ of this distribution:

$$M_n(A) = a\{1 - p[1 - (a/m_n)^{\beta}]/(0.5^{1/n} - 1 + p)\}^{-1/\beta}. \tag{4.69}$$

The second estimator is found as the best unbiased estimator of the parameter A in (4.52). This best unbiased estimator has the following form (see [PLLG]):

$$A = m_n + 1/n\{1 - p + p[1 - (a/m_n)^{\beta}]\}m_n{}^{1+\beta}/[p\beta a^{\beta}]. \tag{4.70}$$

The error range of this estimator is exactly equal to the second term of the right hand side of (4.70). We note that the std is of order $1/n$, since the distribution (4.52) is not regular: its density has a discontinuity at the endpoint $x = A$.

We shall now use (4.69), (4.70) and (4.56), (4.57), and (4.58) to estimate the maximum possible size of seismic disasters and the typical time length in which cumulative losses increase in a nonlinear manner within the framework of the model with sharp truncation (4.52). We will derive estimates separately for our basic catalog 1900–1990 and for the Nelson–Ganze catalog 1900–1979. The most significant difference between the two catalogs consists in the divergence of their respective maximum casualties caused by an individual event: 650,000 in the former case and 240,000 in the latter.

The original data and results of our calculations are presented in Table 4.3. Each calculation was based on the value of β obtained from (4.40) for the respective

Table 4.3 Estimates of maximum possible single loss A and times T_1 (non-linear growth of cumulative sums) and T_2 (linear growth of cumulative sums) for truncation model (4.52)

Equation for parameter A	Estimate of β	Estimate of A, thousands	Error of estimate of A, thousands	Estimate of T_1, years	Estimate of T_2, years
Working catalog					
(3.61)	0.71	1,400 (1,150)		26	32
	0.64	1,100		20	30
	0.78	2,300 (1,600)		40	40
(3.62)	0.71	1,190	550	24	29
	0.64	1,050	400	20	30
	0.78	1,420	750	25	27
Nelson–Ganze catalog					
(3.61)	0.74	505		21	25
	0.63	380		16	26
	0.85	1,005 (650)		25	34
(3.62)	0.74	440	200	19	23
	0.74	380	140	16	26
	0.63	530	290	15	20
Typical values				20–25	25–30

catalog. The result was checked for stability by varying plus/minus the standard deviation (4.40).

We provide the relevant standard deviations for the A-estimates based on equation (4.70). The values of N_1 and N_2 were converted to the lengths of the respective time intervals T_1 and T_2 in years. The basic catalog is long enough to allow A to be found for the annual casualties model. The scatter in A for this model has proved to be significantly higher, which would be natural to ascribe to the smaller statistical database available for this approach. On the whole, the estimates of A based on (4.69) and (4.56) gave rather similar results. Exceptions are the higher estimates of A based on (4.69) for large β. These latter seem to be overestimated, since the approximation (4.43) gradually loses validity as β approaches unity.

As was to be expected, estimates of the maximum possible seismic disaster are strongly dependent on the maximum individual loss m_n. Accordingly, the estimates of A based on the Nelson–Ganze catalog are significantly lower than those for the basic catalog. On the other hand, estimates of typical duration for the nonlinear growth of losses and of the time required for a stable linear growth to settle down have proved to be less wayward. The typical values of T_1 and T_2 were estimated in both cases as 20–25 and 25–30 years, respectively.

Similar results were achieved by modeling the total number of casualties and cumulative economic losses using the bootstrap technique [Ef]. In the bootstrap approach we assume that the theoretical distribution function is identical with the sample distribution function of annual casualties. Then we model the random annual losses for the t-th year ($t = 1, \ldots, 100$), and calculate cumulative sums $\sum_t$. The next step was to find the medians of 1,000 simulations of these $\sum_t$. The results for casualties are shown in Fig. 4.7a. The median of the sums $\sum_t$ increases like $t^{1.53}$ for $t < 20$ years, this being in excellent agreement with the value $\beta = 0.65 = 1/1.53$ given above. Thereupon, for larger times, the growth is decelerating, and for $t > 40$–50 years becomes approximately linear. The least squares estimate of

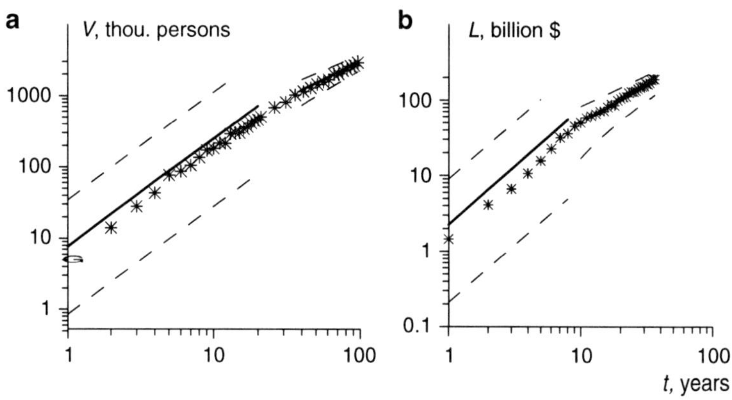

Fig. 4.7 Forecast of casualties (**a**) and economic losses (**b**). *Lines* (solid and dashed) – analytical method, *asterisks* – bootstrap method

the exponent α for $t > 30$ is about 1.06, the value of A being estimated as 700,000–900,000. Similar values of α and half as large estimates of A are obtained by a bootstrap analysis of the Nelson–Ganze catalog for the period 1900–1979.

Figure 4.7 also shows estimates of the cumulative losses using the truncation model (4.52) and the median estimate (4.55), along with 68% confidence intervals. The solid line corresponds either to medians (for $t < 20$ years), or to the normal fit (4.57) (for $t > 40$ years). Dotted lines correspond to 68% confidence intervals for $t < 20$ years and to plus/minus std for $t > 40$ years. The quantiles that provide the 68% confidence level are so chosen, because they correspond to the plus/minus std for the normal law.

Qualitatively similar, though less accurate, results are obtained for the losses. The large scatter of loss values makes it difficult to determine the time since which the catalog can be regarded as homogeneous. Hence one is uncertain about the origin of time to be chosen in the interval 1950–1975. This uncertainty does not matter much, however, because the time interval of interest contains more than 1 year with losses comparable to the maximum \$50 billion, no matter what is the time origin within this interval. Hence we get a more or less stable estimate of the largest observed loss, which is \$50 billion. Figure 4.7b shows results from bootstrap modeling of earthquake losses based on the 1960–1996 data. As the economic loss data are subject to larger errors, we used the annual loss model, which is somewhat cruder. The cumulative losses are increasing roughly like $t^{1.6}$ for intervals shorter than 10 years and like $t^{1.03}$ (nearly linear) for intervals longer than 20 years.

The typical time intervals for earthquake losses T_1 and T_2, as derived from (4.54) and (4.58), are approximately 15 and greater than 30 years, respectively, which is somewhat larger than those given by the bootstrap technique (see Fig. 4.7b). The estimate of maximum possible annual loss relevant to the crossover region is \$150 ± \$100 billion. This large scatter in T_1, T_2 and A must in all probability be related to the poorer statistical database of the loss catalog. We note as well an appreciable distinction between the behaviors of casualties and losses. The former is characterized by a faster growth.

From the above estimates it follows that treatments of worldwide observations of earthquake losses on time intervals below about 20–30 years should take into account nonlinearities in the growth of $\sum_t$. For such intervals of time the sample means (annual, 5-year means, 10-year losses, etc.) are not representative, and total losses $\sum_t$ are comparable with a single (the largest) value. In contrast, the analysis of earthquake casualties on intervals over 50 years can safely rely on conventional methods based on sample means. The crossover point in the statistical behavior of $\sum_t$ is rather uncertain, its location being apparently different for casualties and losses. Apart from errors in raw data and inadequacies in processing methods, this uncertainty may have been due to inadequacies in the model of the truncated Pareto distribution (4.52). One is inclined to think that a gradual-transition model is more realistic. It would be reasonable to recall in this connection that the exponent α in the expression $c{\cdot}t^{\alpha}$ which fits the cumulative sum of casualties and losses approaches unity for time intervals longer than 40 years, while still being slightly above unity ($\alpha = 1.06$ for casualties and $\alpha = 1.03$ for losses). Such behavior of α can be expected in the model of gradual transition for the distribution function. We shall discuss some approaches to gradual transition in Chapter 6.

Calculations similar to the preceding can also be carried out at regional levels. This would require, however, loss data observed during a longer time interval. The latter requirement is related to the fact that the return period of great disasters at the regional level may prove to be much longer than worldwide. This peculiarity is illustrated below using scalar seismic moment data.

The relations derived here permit us to take into account the limitedness of the maximum possible seismic disaster and to evaluate typical losses along with confidence intervals both for the nonlinear domain of growth and for linear growth over long time intervals. These results can be considered as a kind of loss forecast derived on the assumption that the existing statistical patterns will persist later on. How far this assumption is correct will become clearer after discussing the connection between losses and socio-economic conditions in Chapter 7.

The above analysis of earthquake-caused casualties and losses based on the 1900–1999 data suggests the following.

1. The observed nonlinear growth of cumulative casualties and losses can be explained neither by changes in the occurrence of disasters nor by incomplete reporting of casualties for earlier periods. Significant departures from stationarity, largely due to improved detection levels, occur only for smaller events that cause comparatively few casualties and which make up less than 1% of the total number of casualties.
2. The nonlinear growth in cumulative numbers of casualties and losses can largely be explained by the presence of a heavy tail in the loss distributions. In that case the typical cumulative loss is increasing in time like t^{β} with $\beta < 1$. Such nonlinearity can be mistaken for a non-stationarity in the occurrence of disasters.
3. The nonlinear growth of earthquake losses does not last indefinitely. In accordance with world data, its duration is limited by 20–30 years. For larger time intervals further growth becomes ultimately linear, and the maximum size of the loss is saturated. For this reason, for time spans longer than 40–50 years the growth of the losses becomes close to linear.
4. We attempted a forecast of earthquake losses for a future time period, taking into account the abovementioned growth of the losses. The time period for which the forecast would be valid depends on possible changes in socio-economic conditions which can significantly influence the losses. We believe that our forecast is applicable to a time period of about 50 years.

4.3 Main Results

For the truncated Pareto distribution with $\beta < 1$ two kinds of behavior for the cumulative sum are shown to exist: nonlinear and linear. The nonlinear behavior can occur in a stationary model and is explained by the probability of a giant event increasing with time, this providing the main contribution into the cumulative sum. The transition from a nonlinear to a linear behavior is caused by a general limitedness of the observed effect. In the case of earthquake losses, such limitedness can be due to the limitedness that essentially applies to this planet, its population and the total cost of the infrastructure.

The approach to the estimation of cumulative effects we have developed here was applied to the worldwide data on earthquake losses (casualties and economic losses). This type of losses has a complete and accurate data base as compared with the losses due to the other natural disasters. We demonstrated on our data both a nonlinear (for short time spans) and a linear (for longer time spans) character of occurrence of disasters. A forecast of earthquake losses is proposed for about 50 years. Further refinements of the forecast need to take into account the relevant socio-economical situation. This problem will be discussed in Chapter 7.

Chapter 5
The Nonlinear and Linear Modes of Growth of the Cumulative Seismic Moment

In the previous chapter it was shown that the nonlinear mode of growth of cumulative effects takes place in all cases when the distribution function obeys the power-law distribution with $\beta \leq 1$. The distribution of seismic moments obeys this law, and it is this very character of the distribution which causes serious difficulties in seismic risk assessment [PSS, BP, BK, EL, K2, K7, Ki, LTPK, MKP, TLP, Wu, PS1]. The modes of nonlinear and linear increase in cumulative seismic moments are examined below using the world Harvard seismic moment catalog. The catalog includes substantially more data than are available in the case of loss values examined in the previous chapter. The availability of data permits a more comprehensive examination of the distribution behavior in the range of rare large events.

5.1 Nonlinear Mode of Growth of Cumulative Seismic Moment

The distribution of seismic moments is one of the best studied distributions with a heavy tail [PSS, K2, K7, PS1; and many others]. It is easy to see, however, that such a distribution cannot be unlimited. From the general standpoint, a limitation on the occurrence of earthquakes of very high seismic moments and energy is inevitable because of the finiteness of the planet, which precludes the occurrence of especially large earthquakes. Actually, however, a deviation from the power-law distribution with $\beta < 1$ in the tail of the seismic moment distribution is observed for earthquakes that are substantially smaller than a hypothetical event commensurable with the planetary dimensions. Transition from the nonlinear to the linear mode of the growth of cumulative seismic moments is connected with «a downward bend» in the recurrence plot of earthquakes showing in the range of seismic moment values $M \approx 5 \times 10^{27} - 10^{28}$ dyne-cm.

Despite the fact that deviation of the observed seismic moment distribution from the G–R law in the range of high magnitudes is identified fairly definitely, the character of this deviation remains unclear. Because few earthquakes occur in "bend region" and beyond it, the distribution of the greatest events in this range can be described with equal success by several models. The best known modification of the G–R law [K4, K5, K6, KS4] consists in multiplying the power-law

V. Pisarenko and M. Rodkin, *Heavy-Tailed Distributions in Disaster Analysis*, Advances in Natural and Technological Hazards Research 30,
DOI 10.1007/978-90-481-9171-0_5,

distribution of seismic moments by an exponential factor (also referred to as an exponential taper), which leads to a gamma distribution for seismic moments. Most of the other options are represented by the stretched exponential model [LS] and by the power-law model where the parameter β is assumed to be greater than one in the range of large earthquakes [SKKV].

Unlike the majority of the magnitude scales in use, the values of seismic moments do not experience saturation in the range of greatest events [S.I, K2]. The use of the Harvard seismic moment catalog is due to precisely this circumstance. The catalog used below in this chapter (unless specified otherwise) includes seismic moments M for the period from 1977 to 2000. This catalog can be regarded to be complete in the range $M > 4 \times 10^{24}$ dyne-cm, and it is only events in this range which are considered below; there are about 7,000 such events in the catalog. The same limitation is assumed for the synthetic catalogs generated and used below.

Cumulative seismic moments $S(t)$ were calculated for the Harvard catalog as a whole and for a number of individual tectonic regions. The choice of regions was dictated by a certain tradeoff between the requirements of tectonic homogeneity and acceptable statistical completeness. According to their geotectonic settings, the regions are subdivided into groups corresponding to zones dominated by subduction, strike slip movements, continental collision, and mid-ocean ridge zones (MORs). The chosen regions are schematically mapped in Fig. 5.1. The boundaries of regions are based on considerations of tectonic uniformity and were revised in view of uniformity of focal earthquake mechanisms, with A.V. Lander's program allowing visualization of the focal mechanisms being used for this revision. The list of the regions and the numbers of earthquakes with $M > 4 \times 10^{24}$ dyn-cm occurring in these regions are presented in Table 5.1. The time intervals of observation for all regions are the same as that of the whole catalog. Earthquakes with focal depths less 70 km, amounting to 75% of all events, were used in the study of regional seismicity. The sets of earthquakes occurring in depth intervals 70–300, 300–500, and >500 km were analyzed separately.

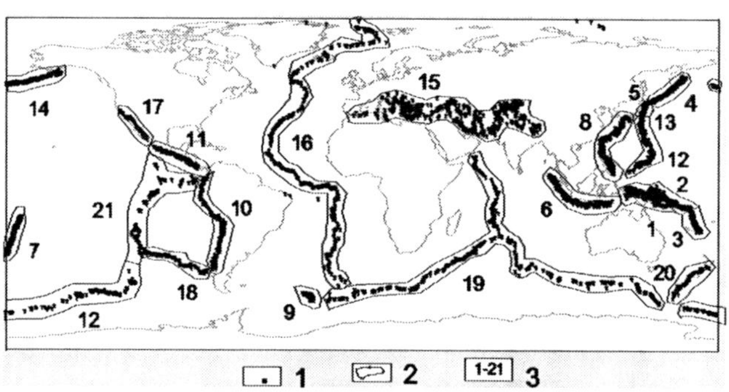

Fig. 5.1 Schematic map showing the regions studied. The points (**1**) are earthquake epicenters ($H < 70$ km), and the *thin black lines* (**2**) are boundaries of the regions. The region numbers (**3**) are enumerated in accordance with Table 5.1

Table 5.1 Characteristics of the used reginal subcatalogs

Region no.	Region	Number of events with $M \geq 4 \times 10^{24}$dyne-cm	Characterization of the region
1	New Guinea	59	Subduction zone
2	Solomon Islands	264	Subduction zone
3	New Hebrides	239	Subduction zone
4	Kamchatka	73	Subduction zone
5	Kurile Islands	118	Subduction zone
6	Sunda Islands	186	Subduction zone
7	Tonga Islands	246	Subduction zone
8	Taiwan	251	Subduction zone
9	South Sandwich Islands	40	Subduction zone
10	South America	176	Subduction zone
11	Mexico	141	Subduction zone
12	Mariana Islands	88	Subduction zone
13	Japan	96	Subduction zone
14	Alaska	151	Subduction zone
15	Alpine belt	222	Collision zone
16	Mid-Atlantic Ridge	147	MOR
17	California	38	Strike slip zone
18	Chile Ridge	62	Strike slip zone, MOR (?)
19	Indian Ocean	193	MOR
20	New Zealand	38	Strike slip zone, subduction zone
21	Pacific Ocean	155	MOR
22	Hypocentral depths of 70–120 km	475	
23	Hypocentral depths of 120–500 km	600	
24	Hypocentral depths >500 km	245	
25	Whole catalog	5,623	

Two approaches are available for the analysis of the time-dependent processes similar to seismicity analysis. In the first approach, aggregated seismic moments over certain time intervals (e.g., 1 year) are analyzed. The second approach consists in analysis of sequences of individual events (the Cramer–Lundberg model [EKM]). In the case of seismicity, both approaches were shown to give similar results. Below, the first approach, which gives a slightly higher effectiveness of computation and robustness, is mostly used. The modes of growth of the cumulative seismic moment values are examined. The Cramer–Lundberg model was used to generate the synthetic catalogs.

Before addressing the analysis of the data, we discuss the possibility of events sharply exceeding (by the order of magnitude) all other events in a given data set. Such anomalous events are considered in [LS] where they are referred to as the king-effects. We show that the king-effect is typical of sets of events that can be described by the unlimited power-law Pareto distribution with $\beta < 1$. Actually, in this case, the distribution function $F(x)$ of the maximum value $X_{max} = max(X_1, \ldots X_K)$ that occurs in a series of K events exceeding a certain threshold a is described as

$$F(x) = [1 - (a/x)^{\beta}]^{K} \tag{5.1}$$

Based on this expression, one can show that, in a series of observations obeying the unlimited Pareto distribution law with $\beta = 1.0, 0.7$, and 0.5, the maximum value in a sample is more than 10 times greater than the median maximum value in approximately 6%, 13%, and 20% of the cases, respectively. For the distribution with $\beta = 0.65$ (the characteristic β value for seismic moments [PS1]), the maximum value more than an order of magnitude in excess will be observed in 15% of the cases. This implies that the value of the maximum element in a sample described by an unlimited Pareto distribution with $\beta < 1$ is highly variable. Indeed, the very presence of such an excessive element (outlier) can be regarded as evidence that the data set under consideration can possibly be described by a Pareto distribution with $\beta < 1$.

The possible occurrence of outliers is taken into account below in the choice of a procedure used in the data analysis. Calculations based upon the Harvard catalog show that cumulative sums of seismic moments $S(t)$ in different regions fluctuate with time very strongly. To obtain stable estimates of the characteristic cumulative seismic moment values, the bootstrap method [Ef] was used. Assuming that the seismic process is stationary in time and using annual sums of seismic moments, 500 random samples $S_i(t)$ of cumulative seismic moment values were calculated for each region in the (0–t) time interval ranging from (0–1) to (0–24) year, $i = 1, 2, \ldots 500$ being the identification number of a random sample. In this way we simulated sets of possible samples involving an increase in the total seismic moment with time. The estimates of the cumulative seismic moments $S(t)$ for each region were obtained by smoothing relevant $S_i(t)$ random samples. As was shown above, in the case of power-law distributions with $\beta \approx 0.65 < 1$, the sample averages are highly variable and tend to increase indefinitely with increasing sample size. Therefore, medians $Med(t)$ of the set of $S_i(t)$ or means $<log\ S(t)>$ (the angle brackets mean the bootstrap averaging) rather than averages of the set of bootstrap samples $S_i(t)$ should be used as robust estimates of the increase of characteristic $S(t)$ values.

Each of these options has advantages of its own. The estimation of the characteristic value by the median of the distribution is more versatile; on the other hand, it was found to be slightly more sensitive to the presence of outliers. The estimation of increasing cumulative sums by the quantity $<log(S(t))>$ is less sensitive to the outliers. However, the standard statistical characteristics of $S(t)$ cannot be directly obtained from estimated mean and variance for $<log(S(t))>$. The quantity $exp(<log(S(t))>)$ is taken as the estimate of $S(t)$. Then, according to the Jensen inequality for convex (in particular, logarithmic) functions, one can assert that this estimate does not exceed the arithmetic mean of the set of bootstrap samples $S(t)$. In our case, the quantity $exp(<log(S(t))>)$ can be used as a smoothed variant of $Med(t)$ values. Both estimates, $Med(t)$ and $exp(<log(S(t))>)$, are used below. As was shown in the previous chapter for a power-law distribution with $\beta << 1$, the characteristic values of the cumulative sums $S(t)$ increase with time t nonlinearly as t^{α}, where $\alpha = 1/\beta > 1$. If the increase in characteristic cumulative $Med(t)$ values is fitted by the relations $log(Med(t)) = \alpha log(t)$ and $<log(S(t))> = \alpha log(t)$, then the case $\alpha > 1$ will correspond to a nonlinear increase in the seismic moments, whereas the case $\alpha \approx 1$ will correspond to the linear mode of increase of characteristic $S(t)$ values with time.

Denote the estimates of the exponent α, obtained for a sample of earthquakes that occurred in a time interval (0, *t*) as $\alpha(t)$. The values of $\alpha(t)$ were determined by least squares applied to earthquakes occurring in different regions in time intervals t varying from (1–4) to (1–24) year for both of the options *Med*(*t*) and *exp*(<*log*(*S* (*t*))>) which represent the change in the characteristic sums *of* $S(t)$ values. Below we discuss the values of $\alpha(t)$ obtained from the relation $<log(S(t))> \cong \alpha(t) \times log(t)$ because they are somewhat more stable. Actually, the regression relation

$$<log(S(t_i))> = a_1 \cdot log(t_i) + a_2, \quad j = 1, 2, \ldots, t \tag{5.2}$$

for a number of t_i values was used to estimate $\alpha(t) = a_1$.

The resulting $\alpha(t)$ values for different regions are presented in Fig. 5.2 where the numbers denote region numbers from Table 5.1.

As expected, the $\alpha(t)$ curves undoubtedly indicate the validity of the relation $\alpha(t) > 1$ and thereby the presence of the effect of a nonlinear increase in cumulative sums of seismic moments $S(t)$ with time. A tendency of decreasing with *t* and of gradual approach of $\alpha(t)$ to $\alpha \approx 1$ is also noteworthy. This effect can be interpreted as a consequence of the transition to the linear mode of $S(t)$ growth with increasing time interval *t*.

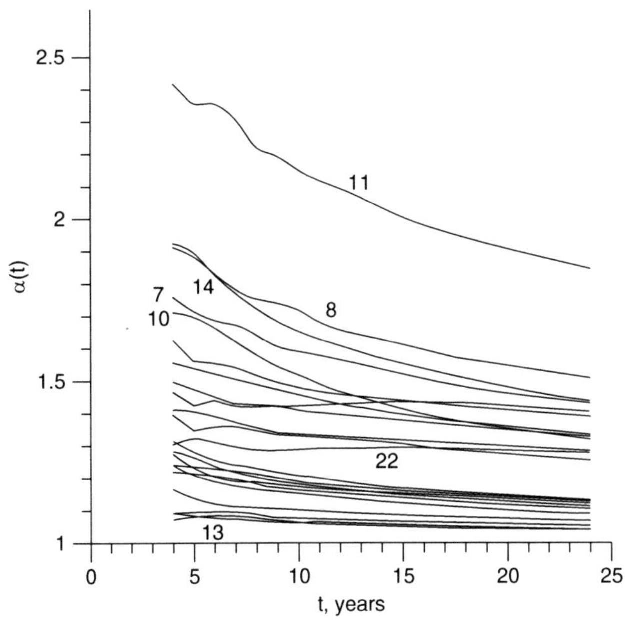

Fig. 5.2 Values of $\alpha(t)$ for various regions for time intervals (0–*t*) year. Wherever possible, the region numbers are shown

Note that the greater $\alpha(t)$ values are observed on average for regions with the lower number of events, and the least values $\alpha(t)$ are obtained for the whole Harvard catalogue. A similar tendency consisting in a more pronounced effect of nonlinear $S(t)$ growth in regions with a lower seismic activity will be shown below using synthetic catalogs. This feature appears to be important for assessing the seismic potential for low seismicity areas with a rare episodic occurrence of large earthquakes.

We will now discuss the nonlinearity of seismicity behavior in greater detail using synthetic catalogs that are presumably similar in their statistical characteristics to the Harvard seismic moments catalog. The duration of synthetic catalogs was set equal to 25 year. The number of events N in the synthetic catalogs was set equal to 25, 50, 75, 100, 200, 500, 1,000, 2,000, 5,000, and 10,000, thereby encompassing the range of the number of events typical of regional subcatalogs and of the whole Harvard catalog. The distribution law of earthquake moments will be simulated by two straight lines (on a log-log scale) that intersect at the point $M = 10^{28}$ dyne-cm. The exponent β for the synthetic catalogs was chosen equal to $\beta = 0.65$ and $\beta = 1.5$ for intervals of seismic moment M values from 4×10^{24} up to 10^{28} dyne-cm and more than 10^{28} dyne-cm, respectively. These β values and the point of intersection $M = 10^{28}$ dyne-cm are chosen according to probable typical values for the real seismic data [PS1].

The purpose of a joint study of synthetic and the observed catalogs is to examine two problems: (1) whether the basic features of regional catalogs agree with the model which differs from the classical Gutenberg–Richter law in the range of large events and (2) whether distinctions between the subcatalogs of different regions can be explained by accidental differences between samples drawn from a uniform general set or regions with different tectonics also differ in the character of seismic moment distribution law. We believe that choice of a model differing from the Gutenberg–Richter law in the range of large events is not very essential here. We remind that, as shown in [PS1], available data do not provide sufficient reasons for preferring any one of a few suggested models of behavior for the tail of the frequency–magnitude relation. The use of any of these models, in particular the model of change in parameter β from $\beta = 0.65$ to $\beta = 1.5$ is, thus, justified.

We now discuss the details in our procedure for generation of synthetic catalogs. Two options of the distribution function are possible. According to the first, the distribution function is

$$F(M) = \begin{cases} 1 - (a/M)^{\beta_1}; & a \le M \le A; \\ 1 - (a/A)^{\beta_1}(A/M)^{\beta_2}; & M \ge A. \end{cases} \tag{5.3}$$

Where $a = 4 \times 10^{24}$ and $A = 10^{28}$ dyne-cm.

From (5.3) it can be seen that the probability density $f(M) = dF(M)/dM$ has a jump at $M = A$. A local maximum of $f(M)$ exists at $M = A + 0$. This maximum does not agree with the classical G–R law, but it can be treated as an analogue of so-called characteristic earthquakes. The hypothesis that postulates the existence of characteristic earthquakes is popular enough [DS, Sc]. According to this model, the distribution of the greatest earthquakes is restricted by a certain maximum hypothetically stemming from the maximum size of the given seismic fault structure. The characteristic earthquake concept was criticized by [Ka1] who showed that the Harvard catalog data do not support this model. However, this model has found support in a number of regional seismicity analyses, and in general the characteristic earthquake model remains popular enough [SS3, and references therein, UPM]. Thus, the model (5.3) can apparently be considered as consistent with some patterns of observed seismicity.

According to the second approach, instead of the distribution function $F(M)$ (5.2), the pdf $f(M)$ can be used, and its continuity at the corner point $M = A$ is required:

$$f(M) = \begin{cases} a_1/M^{1+\beta_1} & a \leq M \leq A, \\ a_2/M^{1+\beta_2}, & M > A, \end{cases} \tag{5.4}$$

where the constants a_1, a_2 are found from the normalization condition and from the continuity of $f(M)$ at $M = A$. In model (5.4) the maximum of pdf $f(M)$ in the range of large events does not exist.

However, for rather short time intervals typical of available earthquake catalogs which contain a few large events $M \approx A$ only, the numerical models (5.3) and (5.4) yield rather similar results. From this one can conclude, that the question of the details in the distribution of extremely large events, in particular, the existence of characteristic earthquakes in concrete seismic regions remains open.

As a matter of fact, regional studies can only reveal a general character of reduced density in the distribution of large earthquakes. From the general requirement that the mean seismic moment and mean seismic energy should be finite one can deduce that beyond some characteristic value M_c the tail $1 - F(M)$ should decrease with M increasing not more slowly than M^{-b}, where $b > 1$. This property is simulated by both (5.3) and (5.4).

Below we use (5.4) for the generation of synthetic catalogs as being more "smooth" and having simpler structure. The return period of earthquakes with the characteristic seismic moment $M = A = 10^{28}$ dyne-cm will be denoted by τ_c. Roughly speaking, for time intervals $t << \tau_c$ the character of the distribution of events with $M > 10^{28}$ dyne-cm is insignificant. The characteristic time τ_c in the model (5.3) depends on the probability of occurrence of strong ($M > A$) events and on the average number of events n per year, and can be roughly estimated as:

$$\tau_c \cong 1/(n \times \phi), \tag{5.5}$$

where $\phi = (\mathrm{a}/A)\beta^1 \approx 0.007$, a probability of occurrence of events with $M > 10^{28}$ dyne-cm.

Now we will describe the results of the numerical simulation. To estimate the statistical characteristics of the synthetic catalogs, 30 random catalogs (samples) were generated for each of the different numbers of events N in the catalog ($N = 25$, 50, 75, 100, 200, 500, 1,000, 2,000, 5,000, and 10,000, as mentioned above). Each synthetic catalog was processed by the bootstrap method, as explained above.

The simulated characteristic dependences $\alpha(t)$ obtained as averaged values of the 30 random samples are presented in Fig. 5.3. The procedure for estimating $\alpha(t)$ estimation for synthetic catalogs was similar to that which was described above and used for regional catalogs. As expected, the values of $\alpha(t)$ decrease with increasing time interval t and increasing number of events N. We begin by considering the second of these dependences.

Speaking loosely, the nonlinear behavior of the cumulative seismic moment is unlikely, if more than one event with the seismic moment $M \geq A = 10^{28}$ dyne-cm occurs with a high probability in unit time (in our case a year). The condition for the occurrence of more than one event with $M > 10^{28}$ dyne-cm in a year has the form $t_c < 1$, which can be rewritten as

$$(N/25) \cdot \phi > 1. \tag{5.6}$$

Remember that $N/25$ is the average number of events per year and $\phi \approx 0.007$ is the probability of events with $M > A$.

We thus can expect that an appreciable effect of nonlinear growth of the cumulative sums of simulated seismic moments in this particular situation should not be observed for catalogs with $N > 4{,}000$ (the number of events with $M > 4 \times 10^{24}$ dyne-cm). This expectation is fulfilled for synthetic catalogs (Fig. 5.3) where the nonlinear effect is clearly visible at $N < 2{,}000$ and becomes insignificant at $N = 5{,}000$ and $N = 10{,}000$. Note that the Harvard catalog which we used contains about

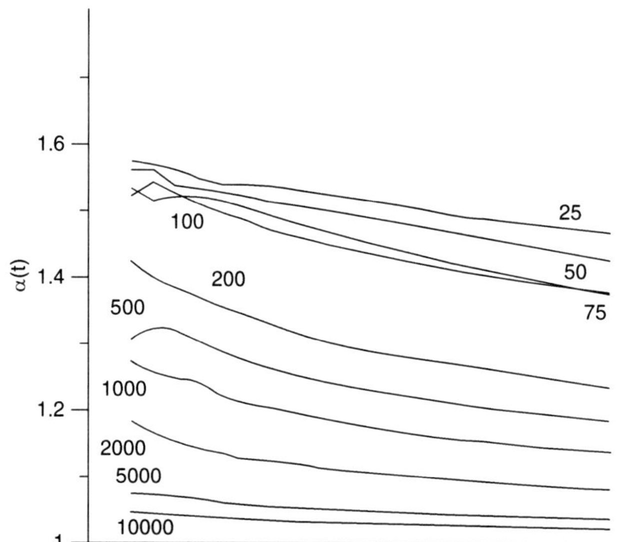

Fig. 5.3 Average values of $\alpha(t)$ for synthetic catalogs for time intervals (0–t) year. The numbers of events in catalogs (N) are shown

5,000 shallow events with $M > 4 \times 10^{24}$ dyne-cm. That means that the nonlinear effect in the whole Harvard catalog is expected to be low and Fig. 5.2 confirms this.

To compare actual and synthetic catalogs, we define a parameter $\alpha_m(N)$ equal to the mean of $\alpha(t)$ values found in the time interval from 4 to 24 year in the set of synthetic catalogs with a given number of events N (or in a given region with a given N value). The subscript m corresponds to different regional catalogs or to different catalogs generated with the help of model (5.4) with a given number of events N. As could be expected, the mean $\alpha_m(N)$ values decrease regularly with N increasing for synthetic catalogs (Fig. 5.4). This tendency is shown in Fig. 5.4 by the solid and dotted lines corresponding to the mean and the standard deviations. The tendency of change of $\alpha_m(N)$ for the actual catalogs is similar, but exhibits an appreciably larger scatter. Actually, as can be seen from Fig. 5.4, the $\alpha_m(N)$ values which pertain to regions differing in seismotectonic settings have their own specific patterns in the $\{\alpha_m(N), lg(N)\}$ plane. Especially wide divergence from the results based on the synthetic catalogs is shown by the $\alpha_m(N)$ values for the MOR zones shown by asterisks in Fig. 5.4. This suggests that the $\alpha_m(N)$ values (which characterize the extent of the non-linearity) depend not only on the number N of events in a region, but also on the seismotectonic situation there. This hypothesis will find support and will be discussed in more detail below.

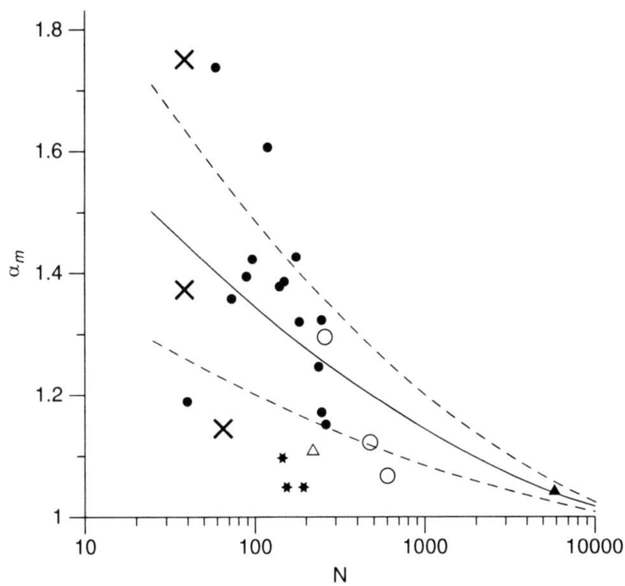

Fig. 5.4 The dependence of α_m on the number of events N in synthetic and actual catalogs. The *solid line* shows the average value of α_m and the *dashed lines* show the scatter for synthetic catalogs. Subduction, MOR, and strike slip zones are marked by *filled circles*, *asterisks*, and *crosses*, respectively. *Filled* and *open triangles* indicate the Alpine belt and the catalog as a whole. *Open circles* indicate deep earthquakes

5.2 Change in the Rate at Which the Cumulative Seismic Moment Increases with Time

The manner in which the nonlinearity pattern in the seismic moment growth is affected by the number of earthquakes in the catalog was examined above. Here we shall discuss how the nonlinearity pattern in the seismic moment growth depends on the time. Special interest attaches to the estimation of the period when the nonlinear growth of cumulative sums $S(t)$ exists for the cases of synthetic and observed catalogues. One clearly sees in Figs. 5.2 and 5.3 a tendency of decreasing $\alpha(t)$ with time t. It is natural to explain this tendency by a gradual decrease in nonlinearity as an ever greater number of seismic moment values becomes available with increasing period of observation. The time interval needed for the transition to the linear mode of accumulation of seismic moments $S(t)$ was crudely estimated above by τ_c (5.5).

The estimates of τ_c obtainable from (5.5) for synthetic catalogs with different numbers of events N in a catalog vary from 1 year up to several 100 years. Numerical experiments with simulated catalogs in time intervals $t > \tau_c$ confirms the transition from the nonlinear to the linear mode of growth of cumulative seismic moments and confirms the validity of τ_c values obtained from (5.5). The validity of identical inferences drawn from observed regional catalogs remains unclear, however, because of an insufficient number of earthquakes and a short period of observation.

To verify the existence of the expected transition from the nonlinear to the linear mode for the growth of cumulative $S(t)$ values, we will consider the tail of the distribution function $1-F(x)$ for the different regions and for the whole Harvard catalog

$$1 - F(x) = (a/x)^{\beta}; \quad x \geq a. \tag{5.7}$$

The distribution (5.7) is equivalent to the Gutenberg–Richter relation and describes the distribution function of seismic moments M. We shall evaluate the values of β_k for k largest events of a given catalogue. Relations (4.13) and (4.14) already used above will be applied to the estimation of β_k [Hi]

$$\beta_k = \left[(1/k) \cdot \sum_{j=1}^{k} log\left(x_j/a\right)\right]^{-1},$$

and

$$Var(\beta_k) = \beta^2/k.$$

In the case under consideration we have $M_1 \geq M_2 \geq \ldots M_k = a$ where M_k is the *k-th* seismic moment value in the given ordered sample. We consider the sequence of β_k estimates as a function of k. The relationship between the β_k and the nonlinearity exponent α can be established as follows. As was shown, if $\beta_k < 1$, then $\alpha = 1/\beta_k > 1$; if $\beta_k > 1$, then $\alpha = 1$. In other words, $S(t)$ increases linearly when $\beta_k > 1$ and nonlinearly when $\beta_k < 1$.

The calculated values of β_k for different regions and for different k are shown in Fig. 5.5. Because of very different numbers of events in different regions, the k values have been normalized by the total number of events N in a given region, $k_{norm} = k/N$. The data for regions with $N \geq 150$ are presented only because the results for regions with smaller numbers of events are hardly representative. As can be seen in Fig. 5.5, the values of β_k are as a rule greater than unity when k_{norm} is small, but decrease and approach the typical value $\beta = 0.65$ with increasing k_{norm}. Hence the distribution of the rare greatest events in the majority of regions certainly differs from the distribution of moderate-sized events. In other words, for time intervals comparable with the recurrence intervals of the greatest events the transition to a (quasi)linear mode of increase for cumulative seismic moments $S(t)$ takes place in most regions.

The data for the MOR zones and for earthquakes with depth of focus greater than 500 km differ from the general pattern. The plots of β_k for the Indian and the Pacific MOR zones are shown by dashed lines in Fig. 5.5 (a similar situation is found for the Mid-Atlantic Ridge, but these data are not presented because of the small $N = 147$ number of events). The β_k values for the MOR zones are substantially larger compared with the other regions and are generally either equal to or greater than unity. This feature is consistent with the well-known difference between the

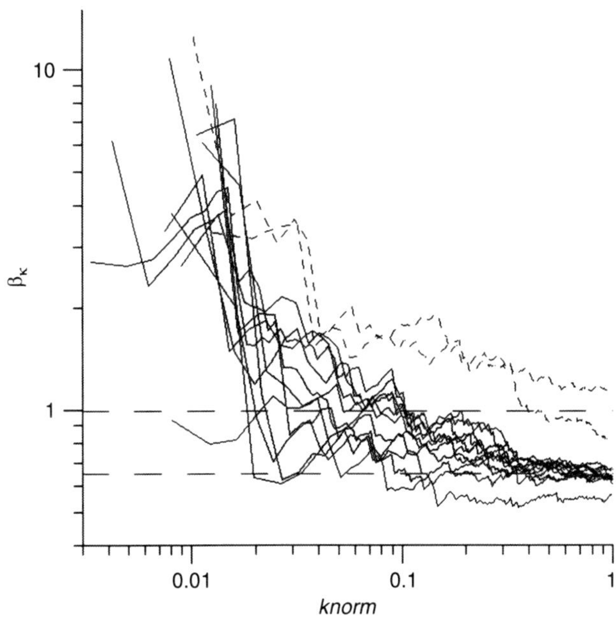

Fig. 5.5 The quantity β_k for various regions as a function of the normalized ranked number of event k_{norm}. The *broken lines* are curves from MOR zones. The *horizontal dashed lines* show the critical and typical values of β, equal to 1 and 0.65, respectively

b-values typical of the MOR zones and other regions [K7, PS1]. The b-values of MOR zones are known to be considerably in excess of the b-values typical of subduction zones and zones of continental collision.

Similarly to the above, where the case of the parameter α was analyzed, we compare results obtained in the analysis of actual and synthetic catalogs; the procedure is to compare the fraction φ of the number of events for which $\beta_k > 1$ with the total number N of events in a given catalog (Fig. 5.6). The scatter in φ values obtained for the simulated catalogs is represented by solid and dotted lines corresponding to the median and the 0.16 and 0.84 quantiles. For synthetic catalogs with the number of events $N < 200$, the confidence intervals for φ are wide and include zero. In other words, the estimates of φ for synthetic catalogs with $N < 200$ fail to establish the transition from the nonlinear $S(t)$ increase to the linear. Note that this numerical finding is quite to be expected, because such N correspond on the average to less than two events with $M > A$.

Comparison of the $\varphi(N)$ pattern obtained for the synthetic catalogs with the results for different regions indicates that the scatter in $\varphi(N)$ for the observed catalogs does not differ from that of the synthetic catalogs, except for the case of the MOR zones. The $\varphi(N)$ points corresponding to MOR zones (marked by stars in Fig. 5.6) appreciably deviate from the model, which agrees well with the well-known conclusion [K7, PS1] as to a special character of the frequency–magnitude relation for the MOR zones.

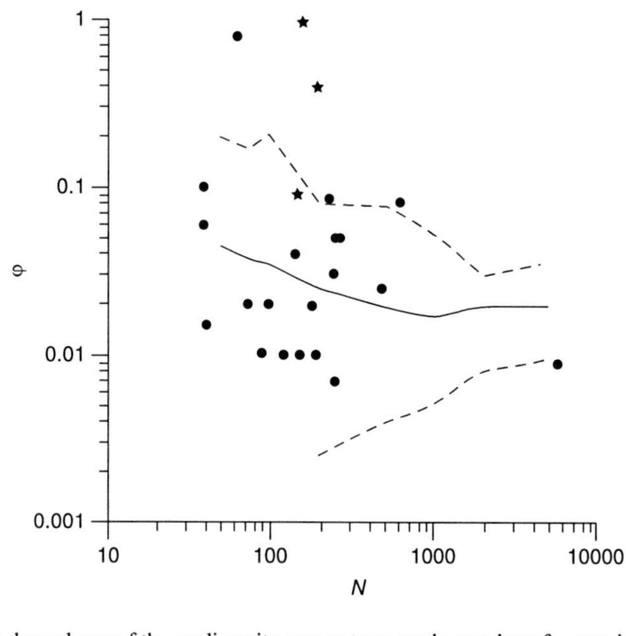

Fig. 5.6 The dependence of the nonlinearity parameter φ on the number of events in a catalog N, for synthetic and actual catalogs. The *solid line* shows the average value of φ, and the *dashed lines* show the scatter for synthetic catalogs. Values of φ in various regions are marked by *filled circles* and those in MOR zones by *stars*

5.3 Characteristic Maximum Earthquake: Definition and Properties

The character of the distribution of rare largest events is the key problem for assessment of risk due to a variety of hazards, including seismic risk. In the previous chapter the model of abrupt cutoff of the G–R law was used because of data shortage. In the case of earthquake death toll and losses no other model could be tested and fruitfully applied. More complex models for the distribution of rare largest events can be examined if seismic moment data are used because of a substantially greater number of events.

Until recently, the seismic risk assessment approaches assumed the G–R law to be unlimited or else to terminate abruptly at a certain maximum possible magnitude M_{max} [We, GMMY, Ki, NMGPFN, KG]. However, recent studies in the distribution of large earthquakes showed both of these assumptions to be inadequate, and some intermediate variant of the distribution law for the largest earthquakes is required.

It was found that the distribution of the largest events deviates from the Gutenberg–Richter law, namely, the number of the largest earthquakes is substantially below what could be expected from the G–R law [K2, PSS, RR1]. Several models have been suggested to describe this deviation. The most popular model [K4, K5, K6, KS4] suggests multiplying the G–R power-law distribution of seismic moments by an exponential factor (also referred to as an exponential taper), which leads to a gamma-distribution for seismic moments. In [LS] the stretched exponential distribution model instead of the classical G–R law was proposed. Lastly, the model of a sharp change of b-value in the G–R law from $\beta < 1$ for $\beta > 1$ in the range of largest earthquakes was proposed [PSS, SKKV, PS1]. This model has been used above in (5.3) and (5.4).

The choice of the most adequate of these models is hardly possible, however. In [PS2] it was shown that because of an insufficient number of the largest earthquakes the character of the deviation of the actual seismic moment distribution from the G–R law cannot be established in detail, and this deviation can thus be described with equal success by different models proposed. A sound choice between these models is possible only when a great amount of additional information on the largest earthquakes has been made available (this requires a long time of observation), or this choice could be based upon convincing reasons of a physical character. Now only the model of change of b-value (of β-value for the case of seismic moment distribution) has a certain physical justification in the form of a possible distinction between the effective dimensions of the earth space available for the occurrence of moderate-sized and largest earthquakes. For the latter case the rupture size essentially exceeds the thickness of the seismic layer of the lithosphere, and thus the effective dimension of space decreases for such events [PSS].

From the aforesaid it can be seen that the problem of determining the distribution law for rare largest earthquakes cannot be thoroughly resolved at present. In this situation it seems reasonable to use a model that would represent a certain generalization of all possible laws. This model is proposed to be the model based upon the notion of the characteristic maximum earthquake M_c and the characteristic return time T_c of $M = M_c$ events. The advantages of this choice will be demonstrated below. Note that the concept of a characteristic maximum event is involved in two of the three models mentioned above. In the model suggested by [K2, K7, KS4] the part of the characteristic maximum event is played by a parameter m_c ("corner magnitude") that marks the beginning of fast decay of the magnitude-–frequency dependence. In the model of change of b-value, the threshold parameter A (see (5.3), (5.4)) represents the characteristic maximum event.

Speaking loosely, the characteristic maximum earthquake M_c is defined as the boundary value between the earthquakes of moderate size described by the Gutenberg–Richter power-law distribution with $\beta < 1$ and the range of the rare largest events described by an unknown density $f(M)$ with a finite expectation.

We shall illustrate this definition by comparing the properties of the actual and synthetic seismic moment distributions in the vicinity of the point $M \cong M_c$. For modeling we shall use the model (5.4) which is specified with the purpose of maintaining the continuity of pdf $f(M)$.

$$f(M) = \begin{cases} C/M^{1+\beta_1} & a \le M \le A; \\ CA^{\beta 2-\beta_1}/M^{1+\beta_2}; & M \ge A. \end{cases} \tag{5.8}$$

In (5.8) $f(M)$ is continuous at point $M = A;$ C is a normalizing constant, it is unimportant and below will be omitted.

We shall use the model already employed above with parameters $\beta_1 = 0.65$ for seismic moments from a $= 4 \times 10^{24}$ up to $A = 10^{28}$dyne-cm and $\beta_2 = 1.5$ for $M > A = 10^{28}$dyne-cm. These parameter values provide a good conformity of the model with the Harvard catalog. The fitted distributions thus obtained with specified points $M = A$ and $M = M_c$ are presented in Fig. 5.7, where the point M_c is defined as the tangent point of the distribution law with the straight line given by $d(log(N))/d\,(log\,(M)) = -1$. Figure 5.7a compares the data from the Harvard catalog (filled dots) and from the synthetic catalog with a large number of events, $N = 10^5$, (grey dots). The large number of events in the synthetic catalog enables us to clarify the position of $M = M_c$, which is found to be a little more to the left from $M = A$. This displacement, however, is rather small and will be neglected in what follows.

In Fig. 5.7b, the distribution obtained for the Harvard catalog is compared with 10 sample synthetic distributions (grey dots), each with the number of events similar to that of the Harvard catalog we used. As can be seen, the distribution of seismic moments from the Harvard catalog agrees rather well with the set of random synthetic catalogs. This agreement provides evidence in favor of model (5.8). Note, however, that the position of the point $M_c \cong A$ is determined with less certainty in these cases because of the smaller number of events.

In order to provide a more specific method of determining M_c we proceed as follows. Consider integral $H(M)$ of pdf $f(M)$ in a moving window $(M/\Delta; M\Delta)$, where $\Delta > 1$:

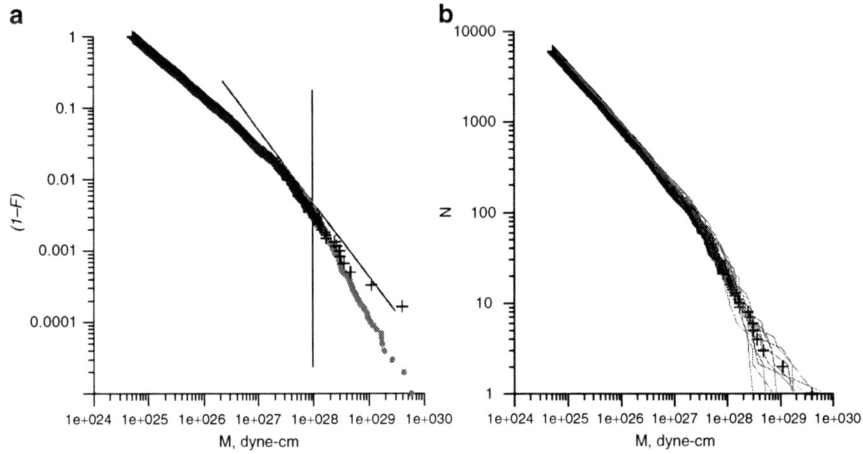

Fig. 5.7 Plots of actual (*crosses*) and synthetic moment distribution (1-*F*). (**a**) *Filled grey circles* show results for the synthetic catalog with 10^5 events and the *lines* show the position of the model crossover point $A = 10^{28}$ dyne-cm and the *straight line* with the unit slope of the recurrence plot. (b) actual distribution and 10 synthetic distributions (*grey lines*) with 7,000 events each are shown

$$H(M) = \int_{M/\Delta}^{M\Delta} f(M)dM \tag{5.9}$$

The domain of integration in (5.8) has the constant width 2·*log*(Δ) on the logarithmic scale *log*(*M*), which also corresponds to a constant window width in the magnitude scale. Substituting *f*(*M*) from (5.8) into (5.9) we get

$$H(M) = \begin{cases} \left[\left(\Delta^{\beta_1} - 1/\Delta^{\beta_1}\right)/\beta_1\right]/M^{\beta_1}; & a + \Delta \leq log(M) \leq A - \Delta; \\ \left[\left(\Delta A^{\beta_2} - 1/\Delta^{\beta_2}\right)/\beta_2\right] \cdot A^{\beta_2 - \beta_1}/M^{\beta_2}; & log(M) \geq A + \Delta. \end{cases} \tag{5.10}$$

Let us denote

$$G(\Delta, \beta) = [(\Delta^{\beta} - 1/\Delta^{\beta})/\beta]. \tag{5.11}$$

Then we have

$$H(M) = \begin{cases} G(\Delta, \beta_1)/M^{\beta_1}; & a \cdot \Delta \leq M \leq A/\Delta; \\ G(\Delta, \beta_2) \cdot A^{\beta_1 - \beta_2} \Big/ M^{\beta_1}; & M \geq A \cdot \Delta. \end{cases} \tag{5.12}$$

Now consider *M*·*H*(*M*), which represents average seismic moment released in the interval (*M*/Δ; *M*·Δ). We take logarithm of *M*·*H*(*M*):

$$log(M \cdot H(M)) \\ = \begin{cases} log(G(\Delta,\beta_1)) - (\beta_1 - 1)log(M); & a \cdot \Delta \leq M \leq A/\Delta; \\ log(G(\Delta,\beta_2)) + log(A^{\beta 2 - \beta 1}) - (\beta_2 - 1)log(M); & M \geq A \cdot \Delta. \end{cases} \quad (5.13)$$

Since $(\beta_1 - 1) < 0$ and $(\beta_2 - 1) > 0$, the function $log(M{\cdot}H(M))$ increases when $M \leq A/\Delta$ and decreases when $M \geq A{\cdot}\Delta$. The value of Δ can be chosen close to 1, then $log(M{\cdot}H(M))$ reaches maximum somewhere near the point $M = A$. Accordingly, the function $M{\cdot}H(M)$ will have its maximum at this point as well. *We define* M_c *(for an arbitrary distribution, not necessarily* (5.10)) *just as the argument providing maximum to* $M{\cdot}H(M)$. For practical purposes we used $\Delta = 0.4$ and $H(M)$ was replaced in (5.10) by its statistical estimate equal to relative number of observations occurred in the interval $(M/\Delta;\ M\Delta)$. We note that such definition provides resulting M_c close to the "corner point" of the tail distribution function, where the condition $d(log(1 - F(M)))/d(log\ (M)) \cong -1$ approximately holds.

We now illustrate this definition numerically. A sample of size $N = 10^5$, obeying the law (5.10) with the parameters $\beta_1 = 0.67$, $\beta_2 = 1.5$, $a = 4 \times 10^{24}$ dyne-cm, $A = 10^{28}$ dyne-cm was generated. The histograms of $M{\cdot}H(M)$ for the Harvard catalog and for the synthetic catalog with 10^5 events are presented in Fig. 5.8a. A definite maximum of $M{\cdot}H(M)$ can be seen in a vicinity of the point $M = A = 10^{28}$ dyne-cm for the synthetic catalog with 10^5 events (solid line). The differences between the positions of the points A, M_c and the maximum of $M{\cdot}H(M)$ are not large, and we can neglect these differences in what follows.

Then 30 series with 7,000 events in each (close to the number of events with $M > 4 \times 10^{24}$ dyne-cm in the Harvard catalog) were drawn at random and histograms of

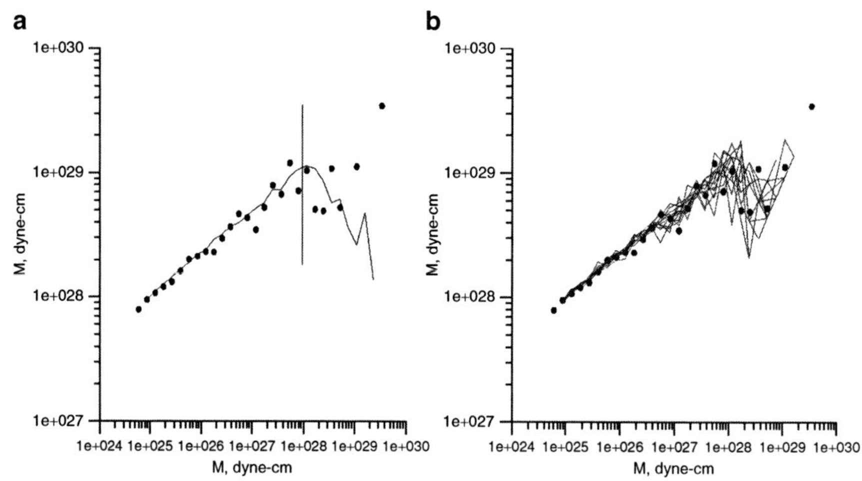

Fig. 5.8 Histograms of released seismic moments for actual (*filled circles*) and synthetic catalogs for equal intervals on $log(M)$ scale. (**a**) synthetic catalog with 10^5 events (*grey line*) and (**b**) for 15 synthetic distributions (in *grey*) with 7,000 events each are shown, *vertical piece* shows position of the model maximum

seismic moments for each series and for the total sample were calculated. In all these cases the total range of M varying from 4×10^{24} to 5×10^{29} dyne-cm was divided into 30 equal log-intervals ($\Delta = 0.4$ for the Harvard catalog).

The resulting $M{\cdot}H(M)$ histograms for the Harvard catalog and for the 15 random samples with 7,000 events each are shown in Fig. 5.8b. It can be seen that the maximum of the $M{\cdot}H(M)$ histogram is somewhat more pronounced for the synthetic samples than for the Harvard catalog. This difference can be caused by differences between the values of M_c in the different regions of the Earth (which turns out to be the case as will be seen below).

We also note that the $M{\cdot}H(M)$ histogram obtained for the Harvard catalog in the range of largest events are actually controlled by the Great Sumatra earthquake of December 26, 2004 ($m_b = 8.9$) and the subsequent large event occurred in the same area on March 28, 2005 ($m_b = 8.2$). These events have produced the two greatest values of the seismic moment in the Harvard catalog. This is why in this particular case we have examined the Harvard catalog in two versions: as usual up to the end of 2000, and up to the end of 2005. In the framework of our model (5.8), the specific form of the histogram corresponding to occurrence of these two largest earthquakes in such a rather short time interval is possible, but is little likely. Only a few of the 15 synthetic samples display a specific form of histogram with an increase in the region of largest events, and none of these synthetic catalogs includes an event similar in size to the Great Sumatra earthquake.

We emphasize that the inference as to the existence of a maximum of the seismic moment (seismic energy) released in the interval $(M/\Delta;\ M{\cdot}\Delta)$ at the point $M = A \cong M_c$ depends weakly on the distribution of the largest earthquakes.

One easily finds that the proposed concept of the characteristic maximum earthquake M_c possesses a number of important advantages. Firstly, it provides appropriate results for all monotonically decreasing probability densities. Since nonmonotone pdf are unlikely to occur, that means that the concept of M_c is applicable to majority of distributions used in practice.

Secondly, the maximum of the released seismic moment (seismic energy) aggregated in windows of equal width in the magnitude scale (analogously in a logarithmic scale of seismic moments) is due to events with seismic moment values close to $M = A \cong M_c$. This property stresses the practical importance of this definition of the characteristic maximum earthquake.

Thirdly (and lastly), the value of M_c does not vary much with varying the lower cutoff magnitude a.

Note for comparison that an alternative introduction of a typical M_{max} value as an exponential index in the Kagan's model [K2, K7] does not possess any of the listed advantages.

For the whole Harvard catalog the value of M_c can be estimated with a sufficient accuracy using the above approaches but at regional scales sufficiently accurate M_c values are difficult to obtain due to the scarcity of large earthquakes. A procedure for robust estimation of M_c at regional scale should be developed. One possible way out of this difficulty is discussed below.

5.4 The Characteristic Maximum Earthquake: Estimation and Application

The concept of the characteristic maximum earthquake with $M = M_c$ and with return period T_c as defined here could be fruitful, if a method for determination of the M_c and T_c applicable at the regional and subregional scale would be developed. We are going to provide such a method using the change in the growth mode of cumulative seismic moments $\exp<\lg(S(t))>$. Such approach provides a more robust determination of T_c and M_c values [R3, R4, RP2, RP4]. The estimation procedure will be examined using both synthetic catalogs and the Harvard catalog. Synthetic catalogs are used to estimate the statistical error inherent to this method.

The synthetic catalog used above, with parameters close to the Harvard catalog, was used again. The duration of synthetic catalogs was put equal to 25 years by analogy with of the used Harvard catalog. Numbers of events in synthetic catalogs were put equal to $N = 25$, 50, 100, 500, 1,000, 3,000 and 10,000 to fit the range of the number of events in different regional catalogs and in the whole Harvard catalog. The distribution (5.8) with values $\beta_1 = 0.65$ and $\beta_2 > 1$ for the ranges of seismic moments (4×10^{24}; 8×10^{27}) and greater than 8×10^{27} dyne-cm, respectively, was used. The poorly known β_2 parameter was examined in different variants: $\beta_2 = 1.1$, 1.3, 1.5, 1.7 and 1.9. To identify the statistical characteristics of the simulated catalogs we generated 30 random samples for each $\{N_i, \beta_2\}$ pair of values. The next step was to use the bootstrap method described above to find the characteristic cumulative sums $S(t)$ for each of these simulated catalogs.

The change in the mode of the cumulative seismic moment $S(t)$ growth from the nonlinear to the linear pattern was captured by the following simple relation:

$$<lgS(t))> = a_2 \cdot lg^2(t) + a_1 \cdot lg(t) + a_0, \tag{5.14}$$

where the parameters a_i ($i = 0, 1, 2$) in (5.14) were estimated by the least squares; broken braces mean the bootstrap averaging. The time interval T_c needed for the passage to the linear mode of $S(t)$ growth was defined as solution of equation

$$d(<lgS(t))>)/d(lg(t)) = 1 \tag{5.15}$$

From (5.14) and (5.15) one has

$$2a_2 \times lg(T_c) + a_1 = 1; \quad T_c = 10^{(1-a_1)/2a_2} \tag{5.16}$$

Relations (5.14), (5.15), and (5.16) provide an estimate of the T_c. If it turns out that T_c is much smaller than the duration of the catalog, then it is desirable to repeat the calculation using (5.14) for a shorter time interval. This recommendation comes from the evident circumstance that relation (5.14) is invalid for time periods much greater than T_c. The procedure described above was tested and found sufficiently robust. The results of this simulation show that, if the number of regression terms in (5.14) is greater than 10, then usually the estimates of a_1, a_2 vary not much. Some unstable results of the use (5.14), (5.15), and (5.16) are obtained, when an especially strong "king-effect" (outlier) occurs in the data set.

Substituting the resulting value of T_c in (5.14), we obtain the estimate of cumulative sum of seismic moments S_{Tc} for T_c years. This S_{Tc} value can be used to evaluate the size of the characteristic maximum seismic moment M_c.This procedure includes two similar steps. In the first step, the value of S_{Tc} is used to estimate the maximum annual value of the seismic moment occured during T_c years. In the second step, the annual maximum value is used to determine the individual maximum seismic moment M_c occurring in the time interval T_c.

Both of these procedures use the interrelation between the maximum term and the sum in a power-law sample with $\beta < 1$ (see Eq. 4.18):

$$S_n = R(n, \beta) Mmax_n,$$

where S_n is the sum of n sample values; $Mmax_n$ is the greatest value in the sample, and $R(n,\beta)$ is a coefficient of the order of unity. The average values of the coefficient $R(n,\beta)$ are tabulated for a set of β,n pairs in Table 4.1.

We now apply the above procedure for estimating T_c and to the analysis of synthetic catalogs. The analysis involves two steps. The first consists in estimating the parameter β_2 (the most uncertain parameter of the model). At the second step the procedure of determining T_c, M_c will be carried out for the case of the best chosen values of β_1 and β_2.

The parameter β_2 in (5.8) describes the moment–frequency relation in the range of the largest events with $M > M_c$. For an estimation of β_2 by the Hill's estimate (4.13) with an acceptable accuracy at least 20–30 events are necessary. But the overwhelming majority of regional catalogs contain few earthquakes of such size. It thus seems impossible to estimate β_2 to sufficient accuracy for the majority of regional catalogs. Fortunately, the results of numerical simulation have shown an absence of substantial dependence of T_c, M_c on β_2. In fact, this result could be expected, because β_2 characterizes the moment–frequency relation in the range $M > M_c$ and for time intervals $t > T_c$. Thus, T_c, M_c can be expected to be tolerant to a variations in β_2.

During the second step we examined the dependence of errors in T_c, M_c on the number of events N in the synthetic catalog. The same set of N values, $N = 25$, 50, 100, 500, 1,000, 3,000, 10,000, was used. The model (5.8) with the parameter values providing the best fit to the Harvard catalog ($\beta_1 = 0.65$, $\beta_2 = 1.5$ and $A = 8 \times 10^{27}$ dyne-cm) was used again. Similarly to the procedure described above, 30 random samples for each N were drawn from the same population and examined. The results of estimation of M_c are presented in Fig. 5.9 where the M_c-median and the 16% and 84% quantiles are shown by solid and dotted lines. In general the accuracy of these values of M_c is satisfactory. An expected increase in the accuracy with increasing number of events N is clearly visible.

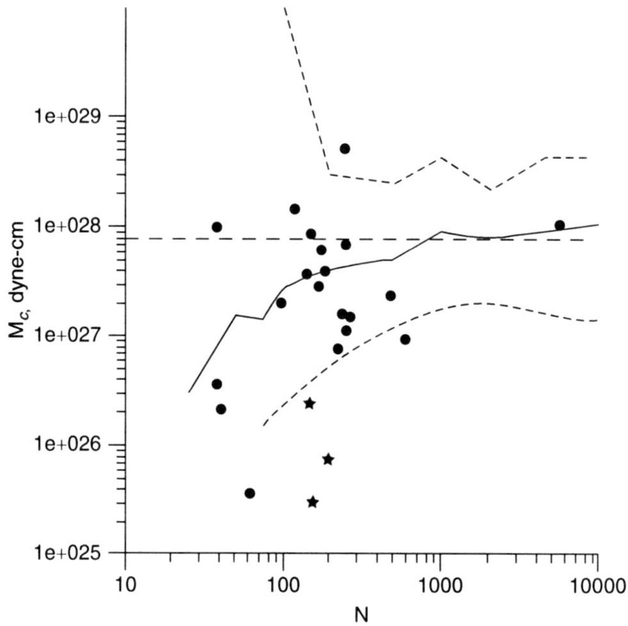

Fig. 5.9 The dependence of M_c on the number of events N in synthetic and actual catalogs. The *solid line* is the mean and the *dashed lines* show the scatter of M_c for synthetic catalogs. MOR zones and other regions are marked by *asterisks* and *filled circles*, respectively, *horizontal broken line* - model Mc value

After this examination of the accuracy for the ideal case of synthetic catalogs we proceed to a similar examination for the actual regional data. The procedure of M_c estimation is identical with that used for synthetic catalogs. The results obtained are rather similar also. As can be seen from Fig. 5.9, the estimates for different regions (the MOR zones excluded) agree quite well with the results obtained for synthetic catalogs. The results obtained for the mid-oceanic rift (MOR) zones (shown in Fig. 5.9 by asterisks) disagree definitely with the pattern obtained for the synthetic catalog. This disagreement is quite expected because the parameters used in (5.11) do not correspond to those describing the moment-–frequency relation typical of the MOR zones. The typical β-values for the MOR zones exceed unity (see Figs. 5.5 and 5.6), and thus the fraction of earthquakes with $M > M_c$ is rather high; accordingly, the values of M_c are low, which can be seen in Fig. 5.9.

The calculated parameter values for the regional catalogs are presented in Table 5.2 (some parameters of the regions were given earlier in Table 5.1 and Fig. 5.1). For convenience of comparison the main tectonic characteristics of the respective subduction zones (from [Ja]) are also given in Table 5.2. Note that in a few regions M_c values could not be estimated robustly because of the "king-effect" of outliers.

Table 5.2 Seismotectonic parameters for different regions

Region number from Table 5.1	Regional parameters of seismicity[a]									
	1	2	3	4	5	6	7	8	9	10
1	1.13	−1.24	0.005	4.3	50	1	3.43	0.3	–	–
2	0.96	−0.64	0.05	12	50	4	4.88	0.31	0.15	23
3	0.71	−0.85	0.02	8.8	52	1	4.43	0.44	0.16	16
4	3.11	−1.0	0.02	8.8	90	5	2.98	0.15	–	–
5	3.96	−0.58	0.008	8.7	119	5	3.33	1.09	1.45	63
6	1.7	−0.62	0.01	8.2	88	5	3.38	0.58	0.4	43
7	1.24	−0.69	0.006	7.5	117	1	2.94	0.88	0.68	41
8	1.09	−0.6	0.05	4.6	80	3	3.64	0.32	0.12	17
9	1.04	−0.7	0.12	0.9	49	1	2.45	0.02	0.02	53
10	1.14	−0.52	0.02	10	38	7	3.62	1.28	0.62	22
11	0.55	−0.45	0.04	7.2	17	6	4.27	0.87	0.36	16
12	1.41	−0.84	0.005	7.6	94	4	2.52	0.12	–	–
13	2.09	−0.71	0.015	9.9	67	6	3.87	0.38	0.2	37
14	1.29	−0.62	0.01	6.3	49	6	3.57	0.41	0.86	90
15	1.2	−0.62	0.05	–	–	–	3.11	0.17	0.08	25
16	0.66	−0.77	0.04	–	–	–	2.38	0.05	0.02	26
17	0.66	−0.61	0.06	–	–	–	3.55	0.06	0.04	30
18	0.34	−0.92	0.28	–	–	–	3.31	0.01	0.004	22
19	0.54	−0.8	0.15	–	–	–	3.3	0.02	0.01	14
20	1.01	−0.45	0.02	–	–	–	5.67	0.33	1.0	84
21	0.27	−1.1	1.0	–	–	–	3.23	0.02	0.003	9
22	1.72	−0.69	0.04	–	–	–	5.62	0.44	0.23	21
23	1.23	−0.61	0.07	–	–	–	6.46	0.26	0.09	14
24	0.9	−0.58	0.07	–	–	–	5.5	3.04	1.05	27

[a](1) Median apparent stress σ_a, bar; (2) β in the linear part of the moment–frequency relation; (3) φ; (4) subduction rate, cm/year; (5) plate age, 10^6 years; (6) "stress class"; (7) median of seismic moments, $\times 10^{24}$ dyne-cm; (8) M_{3max}, $\times 10^{28}$ dyne-cm; (9) M_c, $\times 10^{28}$ dyne-cm; (10) T_c, years

Some of the results can be treated as consequences of the model used. Thus, the tendency of T_c increasing with decreasing number of events N in the region was expected, and this tendency is observed both for actual and for synthetic catalogs. The correlation coefficient r between T_c and N for actual catalogs was $r = -0.25$. The rather small value of the correlation coefficient is due both to the errors in T_c and to the expected difference in T_c values in different regions because of different tectonics (note that in synthetic catalogs both T_c and M_c values were fixed).

We now compare the resulting M_c values for different regions. Firstly, we ask whether the M_c estimates correlate with the observed maximum seismic moments. But, as has been emphasized, the maximum value in a sample drawn from a power-law distribution with $\beta < 1$ is highly unstable. As a more robust characteristic, the geometric mean of the three largest seismic moments values was used: $M_{3max} = (M_1 \cdot M_2 \cdot M_3)^{1/3}$. The M_{3max} values were compared with the M_c estimates. One can see rather close correlation (Fig. 5.10).

Compare now other characteristics of seismicity for the different regions. Figure. 5.11a compares the M_c and the β characterizing the linear part of the moment–frequency relation. A certain correlation between these characteristics

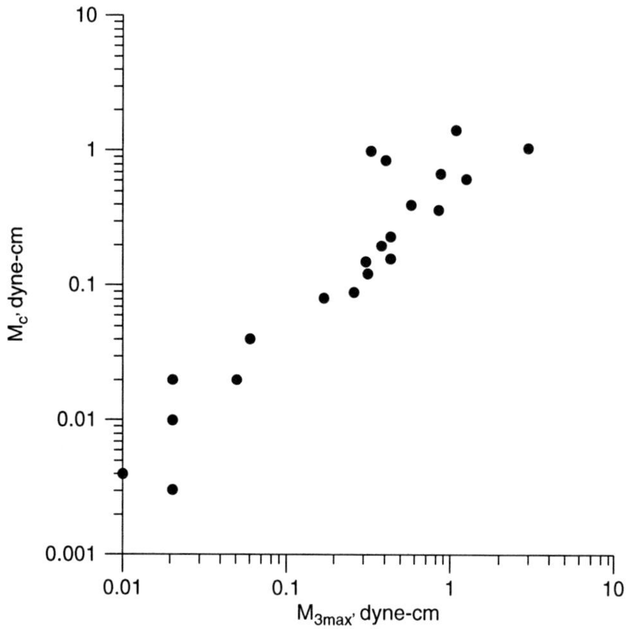

Fig. 5.10 Correlation of M_c and M_{3max} in different regions

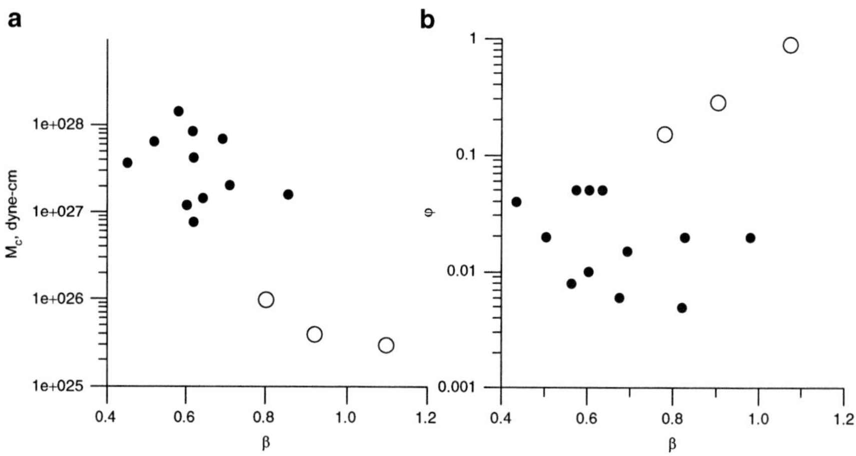

Fig. 5.11 The dependence of β in the linear part of the moment–frequency relation and M_c (**a**) and φ (**b**) values. MOR zones and other regions are marked by *open circles* and *dots*, respectively

can be seen. In other words, a correlation exists between the distribution pattern of moderate-sized earthquakes described by the β-value and the typical size of large earthquakes characterized by the M_c. Figure. 5.11b shows the β-values for different regions versus the φ-parameter; remember that the φ-parameter characterizes the fraction of events with seismic moments $M > M_c$. The well-pronounced tendency of increasing φ with the β-value was quite expected.

In both Figs. 5.10 and 5.11a, b the points corresponding to the MOR zones and to the other regions occupy quite different positions, thus corroborating the well-known fact of a significant difference in the character of seismicity between the MOR zones and the other seismic regions. However, the character of the above correlations shows that the MOR zones do not form a special separate group of regions; rather, the MOR zones occupy an extreme position within the common continuous tendency of changing seismicity parameters which is observed for all seismic regions no matter what is their tectonics.

5.5 The Seismic Moment–Frequency Relation: Universal?

The above analysis shows that the variability of M_c among regions is stronger than it takes place for the synthetic catalogs with similar number of events. That means that M_c depends on the tectonic setting, which is different in different regions. If so, the well-known hypothesis of universality (see [K2, K7, K8, K9, GP]) of the relation between seismic moments and frequency of occurrence is not confirmed.

The hypothesis of universality should perhaps be discussed in more detail. It follows from the practice of seismicity research that a more complete analysis of historical, geological and tectonic data generally results in a higher estimate of seismic potential for the region of study. This tendency is typical both of active seismic regions, such as, the Caucasus (see [Ni, RN]), and of low seismicity regions. Even regions that were previously believed to be of low seismic activity were sometimes hit by large earthquakes. As a result of a detailed research followed such large earthquakes, the traces of past large earthquakes, paleoseismic deformation and historical evidence come to light, showing that similar large events had occurred there in the past. One of the last examples of such revision followed the Neftegorsk 28 May, 1995, $M_s = 7.2$ damaging earthquake (Sakhalin Island, Russia) described in detail in [RRB, Str]. A similar tendency is observed in other countries. Thus, the Tangshan earthquake (July 28, 1976, China, $M_s = 7.8$), the most disastrous in the last century, occurred in an area that had been believed to be of low seismic activity. The maximum seismic intensity expected to occur there was equal to VI, whereas the observed intensity exceeded IX.

The possibility of occurrence of large earthquakes in areas believed to be before of low seismic activity is explained by the short period of observation and by long repeat periods of large earthquakes, which can be much longer than the period of registration. Actually, the recurrence time of the largest earthquakes in a number of regions of moderate and low seismic activity was estimated to be several thousand years [GGG, SGZLB].

The estimation of probability of occurrence for rare large earthquakes is traditionally based on the Gutenberg–Richter frequency–magnitude relation, the procedure being to estimate the probability of rare large earthquakes from the probability of frequent small earthquakes obtained by seismic monitoring. Using this approach one gets the result that the possibility of occurrence for large earthquakes in an area where no small earthquakes have been recorded is extremely low. However, the character of the frequency–magnitude law in the range of rare large earthquakes is poorly known (see the detailed discussion in Chapter 6), so several versions of the tail distribution were suggested. Among others there is the model of the so-called characteristic earthquakes (a relative increase in the probability of large earthquakes in some particular regions, see [DS, Sc, UPM]).

The most adequate characteristic of earthquake size is the seismic moment, where the log scale has no saturation effect for the largest events (see [So1, K2]. In view of this circumstance, and also because of a high uniformity of worldwide data available in the Harvard seismic moment catalog since 1976, it is this very catalog which is used as a rule for analyses of global seismicity. Note also that no evidence of occurrence of characteristic earthquakes in the global Harvard catalog was found [Ka1].The hypothesis of the universal moment–frequency relation, put forward by Y. [K2, K7], consists in the following: the form of the moment–frequency relation is universal, these relations in different regions differ only by the rate of seismicity, i.e. by average number of events (of some particular size) per the unit time. In other words, all moment–frequency (or magnitude–frequency) curves, corresponding to different regions, are similar and differ only by a factor. Later on, this hypothesis has been refined, and the evident difference in seismicity between the MOR zones and the continental and subcontinental areas was accepted and taken into account; the moment–frequency relation throughout all continental areas is assumed to be universal.

It is obvious that confirmation or rejection of the hypothesis of universality would be very important both in theory and in practice of seismic hazard assessment. If the hypothesis of universality is confirmed, and the form of frequency–magnitude relation is well established on the global level, the regional seismic hazard would be completely controlled by a single parameter, viz., the rate of earthquakes per unit time. Otherwise, a much more complex model should be examined, because possible regional and subregional differences in the moment–frequency relation should be taken into account.

The hypothesis of universality will be discussed below. We shall use, unlike the earlier works on this subject [Rk1, Ja], the parameters measuring the seismicity level, that do not depend on the time interval, covered by the catalog in question. The characteristic maximum event M_c is an intensive parameter not depending (in principle) on the time span of the catalog registered in a particular region. On the contrary, the observed in a time span T maximum seismic moment is a parameter having tendency to grow with increasing time of registration T. From the detected above large variability in M_c from region to region it follows, that the hypothesis of universality is not confirmed.

Further testing of universality is based on the comparison of the intensive parameters of seismicity in different regions with the tectonic characteristics of these regions. We used the following seismicity parameters: M_c, T_c, β-value, and φ.

The subduction zones were characterized by: stress class, subduction rate, and plate age (all characteristics were taken from [Ja]). Besides, the mean apparent stress σ_a calculated from the Harvard catalog was taken into account. The apparent stress values were calculated from the well-known relation [Ab1]

$$\sigma_a = \mu E/M, \tag{5.17}$$

where μ is the shear modulus and E is the seismic energy calculated by the standard way from the m_b magnitude [So1]. The shear modulus μ was assumed to depend on depth according with [Bu2].

The correlation coefficients between the above parameters are presented in Table 5.3 where the bold figures mark significant correlations exceeding the confidence level 90%. In some cases we used the log-transform instead of the linear scale. For example, subduction rate is given on the linear scale, while the characteristic maximum seismic moment M_c is given in the log scale. In several cases both of these variants are reasonable, and in these cases the result corresponding to a higher correlation is presented in the table. Some of the correlations should be expected from formal reasons or from the physical meaning of the parameters under discussion. The actual existence of such correlations supports the validity of our approach.

It is of the interest to see whether there are correlations between intensive characteristics of seismicity (parameters 1, 2, 3, 7, 9, and 10) and tectonic characteristics (parameters 4, 5, and 6). In fact, there are a few significant correlations of this kind. Such correlations provide evidence against the hypothesis of universality.

Interpretation of the identified correlations deserves a special research. Some aspects of such an interpretation are discussed in the following chapter. Here we note only the a priori plausibility of the obtained correlation of M_c with the subduction rate (rather weak, however).

Table 5.3 Correlation coefficients for tectonic and seismicity parameters for different regions[a]

Parameters	Parameters									
	1	2	3	4	5	6	7	8	9	10
1	1									
2	0.1	1								
3	**−0.6**	**−0.44**	1							
4	0.24	0.2	**−0.52**	1						
5	**0.72**	−0.12	−0.47	0.12	1					
6	0.3	**0.52**	−.03	**0.54**	−0.16	1				
7	−0.07	0.39	−0.12	**0.56**	−0.5	0.24	1			
8	**0.54**	0.4	**−0.62**	**0.73**	0.05	0.4	**0.48**	1		
9	**0.73**	**0.74**	**−0.84**	**0.69**	0.5	**0.51**	0.41	**0.94**	1	
10	**0.65**	**0.43**	**−0.64**	−0.29	0.44	0.14	−0.13	0.25	**0.67**	1

[a](1) Median apparent stress σ_a, bar; (2) β in the linear part of the moment–frequency relation; (3) φ; (4) subduction rate, cm/year; (5) plate age, 10^6 years; (6) "stress class"; (7) median of seismic moments, $\times 10^{24}$ dyne-cm; (8) M_{3max}, $\times 10^{28}$ dyne-cm; (9) M_c, $\times 10^{28}$ dyne-cm; (10) T_c, years

Summarizing, one can conclude that the results of the above analysis testify definitely against the hypothesis of universality of the moment–frequency relation. This conclusion will receive further support in the next chapter where considerable differences between the maximum possible magnitudes will be shown to exist for a number of continental regions.

5.6 Nonlinear Mode of Growth of Cumulative Seismotectonic Deformation

The nonlinearity of the growth of cumulative seismic moments should cause a similar effect in the growth of cumulative seismotectonic deformation. We shall analyze this effect. The amplitude $U(t)$ of seismotectonic deformation occurring along a fault is expected to be related to the characteristic cumulative seismic moment $S(t)$ of earthquakes occurring at this fault in a given time interval via the well known relation [Br, Kas]

$$U(t) = S(t)/\mu A_o, \tag{5.18}$$

where μ is the shear modulus, and A_o the area of the surface of the megafault on which the movement occurs. Within the framework of this approach, all seismic events that occur in a given region are treated as episodes of some uniform dislocation process occurring along a surface A_o during a given time interval. This assumption is not quite realistic, but it provides a crude modeling of seismotectonic deformation occurring along major strike slip zones, subduction zones included. Regions 1–14 from Table 5.1 (subduction zones) and regions 17, 18 and 20 (strike slip zones) could be considered as regions with this type of deformation. However, sufficient number of earthquakes needed for a reliable statistical analysis there exist only for the subduction zones. Therefore below our study will be restricted to the first 14 regions from Table 5.1.

As rough estimates of the cumulative seismic moment we shall use the calculated above regional values of $exp(<log(S\ (t))>)$. Following [Kas], the shear module μ will be set equal to 3×10^{11} dyne/cm^2. The megafault surface A_o was estimated from the length of the corresponding section of the deep-sea trench and the length of the dipping subduction zone from the ground surface down to 70 km depth. These calculations were based on information from [Pl, Ja].

The resulting typical rates of seismotectonic motion for the subduction zones are given in Fig. 5.12 for time interval of 1–24 years. In accordance with the nonlinear increase of cumulative seismic moments $S(t)$, there is also the effect of a nonlinear increase in the rate of seismotectonic deformation. In a number of regions the rate of seismotectonic deformation becomes rather stable toward the end of the averaging time (15–20 years); but in other regions the rate continues to increase throughout the entire accessible interval of registration.

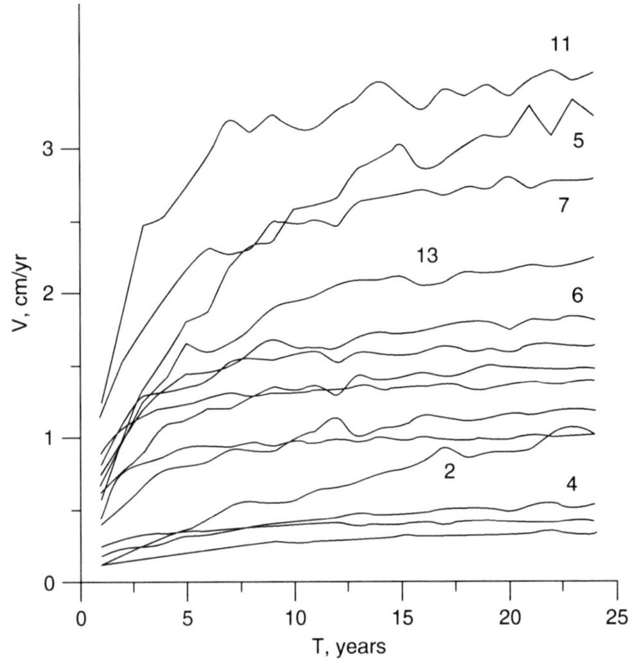

Fig. 5.12 The dependence of estimated rates of seismotectonic deformation V on T (time interval of averaging). Wherever possible, the region numbers are shown

The maximum rate values corresponding to the greatest available interval of averaging vary from 0.3–0.5 to 3–3.5 cm/year. Comparing these values with the rates of seismotectonic deformation obtained by other methods, we conclude that the approach used here leads to noticeably smaller rates than those obtained by other methods. This difference has two main explanations. Firstly, the seismotectonic rates are, as a rule, less than the rates obtained from geodetic and other similar measurements because of a large contribution due to slow aseismic creep deformation in the total deformation. Besides, it should be taken into account that the values of $exp(<log\ (S\ (t))>)$ we used are geometric means, whereas the usual procedures of averaging give arithmetic means. But it is well known that arithmetic means in most cases noticeably exceed geometric means, and are never lower. But the arithmetic means are much more sensitive to the outliers. The advantage of our approach consists in the robustness of the results revealing tendencies in the seismotectonic deformation on the background of high fluctuations of the seismic process.

The rates of seismotectonic deformation obtained for the 24-year interval of averaging were compared with the corresponding subduction rates derived for longer time intervals [Ja]. The correlation coefficient r between these data sets was found to be 0.43; thus, a certain interrelationship exists between these two characteristics, even though rather weak. The constant of proportionality between the derived rates of seismotectonic deformation and the subduction velocities varies for different subduction zones from 0.05 to 0.45.

The nonlinear and the linear modes of growth of cumulative seismic and seismotectonic effects are especially important in the analysis of long-term seismic hazard in low seismicity areas. The neglect of this effect can lead to a serious underestimation of earthquake hazard when time span of the analysis is not sufficient [R3, R4]. This tendency is closely connected with the effect of rare accidental occurrence of large earthquakes in areas that had been thought to have a low level of seismicity. The correct analysis of nonlinear and linear modes of growth of cumulative seismic effects can provide some upper bound on the probability of the rare largest earthquakes. This estimation "from above" will be nevertheless noticeably below what would have been derived by extrapolation of the usual Gutenberg–Richter magnitude–frequency relation. A proper consideration of this point could be especially valuable for the assessment of earthquake hazard in the case of radioactive and chemical waste depositories that should be kept in safety for tens of thousands of years [MRT]. The similar situation takes place with the volcanic hazards [HS].

5.7 Main Results

Data on seismic moments are much more numerous than data on losses from natural disasters. This information has allowed to investigate the tail of the seismic moment distribution in more detail than it has been made in the previous chapter for the earthquake loss data. Accordingly, instead of the model in which the tail of the distribution is truncated, there is a possibility to elaborate a more adequate model for the tail distribution. A few such models were suggested by different authors. However, the available seismic information is insufficient to select one "true" model among all.

In this situation a simple model was suggested, intended to capture the main features of other models. This model is based upon the crossover point M_c that divides the range of seismic moment into two parts. The first part corresponds to the classical Gutenberg–Richter power-law distribution with parameter $\beta < 1$. The second part corresponds to the unknown distribution law of large earthquakes with a finite mean. This part of the distribution can be modeled by the power law with $\beta > 1$. The reasonableness of such definition of the characteristic maximum seismic moment M_c is supported by the following remark. The maximum energy release values takes place around $M = M_c$ if the energy is summed in windows of constant log-length.

The robust procedure for the estimation of the characteristic maximum seismic moment M_c corresponding to the crossover point and the recurrence time T_c was proposed and tested. The parameters M_c, T_c have been determined for a number of different regions and were shown to depend on the tectonic characteristics of the regions. These results testify that the hypothesis of universality of the seismic moment–frequency relation is not confirmed.

The nonlinear mode of increase of the cumulative seismic moments (and seismic energy) seriously complicates the assessment of the earthquake hazard, especially in low seismicity areas where the period of such nonlinear mode can last as long as hundreds and thousands of years. This problem can be especially acute for the cases where earthquake hazard has to be assessed for critical facilities that are envisaged for very long time periods and located in low seismicity areas.

Chapter 6
Estimating the Uppermost Tail of a Distribution

This chapter is devoted to the study of earthquake size distributions in the uppermost range of extremely rare events. We suggest a new method for statistical estimation of the tail distribution. The method is based on the two main theorems in the theory of extreme values and on the derived duality between the Generalized Pareto Distribution (GPD) and the Generalized Extreme Value distribution (GEV). Particular attention is paid to the adequacy of the widely used parameter M_{max}. This parameter, as will be demonstrated, is very unstable, in particular, in highly seismic regions. Instead of this unstable parameter we suggest a more stable and robust characteristic: the quantile of confidence level q of the random maximum magnitude in a future time interval T, which we shall call the T-maximum. We shall set down methods for practical estimation of such quantiles. Also, some related problems of seismic risk assessment are discussed.

6.1 The Problem of Evaluation of the "Maximum Possible" Earthquake M_{max}

The Gutenberg–Richter (G–R) relation has been the subject of extensive research (see, e.g. [BK, CFL, K1, K3, K7, K8, K9, KS1, KKK, MIMR, OK, PSSR, PS2, SKKV, U, Wu]). One can summarize the present situation by saying that, for small and moderate magnitudes, and for large space-time volumes, the Gutenberg–Richter relation is valid to high accuracy. However, for the largest magnitudes, some more or less significant deviations from the G–R have been documented (see, e.g., [PS2] and references therein). The study of such magnitudes is hampered by the insufficient number of large earthquakes. Inevitably, existing models of the deviations from the G–R and the numerous proposals to modify it for large magnitudes suffer from a large statistical uncertainty. As a consequence, the problem of finding an adequate description of the tail of the magnitude distribution cannot be considered as definitely settled. One of the best known modifications of the G–R

V. Pisarenko and M. Rodkin, *Heavy-Tailed Distributions in Disaster Analysis*, Advances in Natural and Technological Hazards Research 30,
DOI 10.1007/978-90-481-9171-0_6,

([K4, KS4]) consists in multiplying the power-law distribution of scalar seismic moments (which, as we recalled above, corresponds to the G–R exponential distribution of magnitudes) by an exponential factor (also referred to as an exponential taper), which leads to a gamma distribution for seismic moments. The index in the exponential taper is often referred to as the "corner" moment, as it constitutes the typical value at which the distribution departs significantly downward from the pure G–R law. The corner moment does not represent the maximum possible size, and larger moments are authorized in this model, albeit with exponentially smaller probability. This exponential taper to the power law of seismic moments constitutes one among several attempts to take into account the presence of a downward bend observed in the empirical distribution of earthquake size for the largest moments (or magnitudes), see [BK] for the use of the Kagan model to determine the corner magnitude for seven different tectonic regions.

Rather than introducing a "soft" truncation of the G–R law, a different class of models of $F(x)$ assumes that the G–R law holds as far as the maximum magnitude M, beyond which no earthquakes can occur [CFL, Da, MIMR, P2, PLLG]:

$$F(x) = \begin{cases} 0; & x < m; \\ [10^{-\beta m} - 10^{-\beta x}]/[10^{-\beta m} - 10^{-\beta M}]; & m \leqslant x \leqslant M; \\ 1; & x > M. \end{cases} \tag{6.1}$$

The parameter M represents the *maximum possible earthquake size:* $M = M_{max}$. This parameter plays a very important role in seismic risk assessment and in seismic hazard mitigation (see, e.g., [BP, PLLG, KG, KLG]). It should be noted that the truncated G–R (6.1) ensures that the mean seismic energy is finite, whereas the G–R in its unlimited form corresponds to a regime with infinite seismic energy, which is, of course, an undesirable property of the model (the exponential taper in the Kagan model of seismic moment distribution also ensures the finiteness of the mean seismic moment). The parameter M_{max} is very suitable for engineers and insurers: having a reliable estimate of M_{max}, it is comparatively easy to take adequate decisions on the standard building code or about insurance policy. As a consequence, the truncated G–R (6.1) has become very popular. Unfortunately, all attempts so far for a reliable statistical estimation of M_{max} did not give satisfactory results. The statistical scatter of its estimates and their reliability are far from being satisfactory. Attempts to attribute a maximum magnitude to individual faults rather than to regions suffer from the same problems and in addition face the fact that many large earthquakes involve several faults or fault segments which are difficult, if not altogether impossible, to identify in advance [W].

Below we explain and illustrate the intrinsically unstable properties of the determination of M_{max}. Besides, M_{max} has the following undesirable features.

1. M_{max} is ill-defined. It does not contain ***the time interval (or, time scale)*** over which it is valid. Suppose we are supplied with a very good estimate of M_{max}. Does this value apply to the whole geological history (4.5×10^9 years) or to the

last geological time period since the Gondwana accretion (200×10^6 years), or else to the last 200 years? This question, we believe, should be very important for practical applications, but M_{max} does not answer it at all. A way to circumvent this problem is to define M_{max} as the magnitude of an event with exceedingly low probability over a finite time period, say the next 30 years. This last definition is more in line with the needs of insurance and risk policy.

2. It is a fact that the maximum earthquake magnitude should be finite because the Earth is finite. One can however use the "small increment" ε-reasoning (where ε is an arbitrary small positive number) to argue that "if a magnitude, say M^*, is the maximum possible, then why is magnitude $(M^* + \varepsilon)$ impossible?" It is difficult to find a convincing answer to this question.
3. Finally, the parameter M_{max} is intrinsically unstable from the statistical point of view. We are going to demonstrate this fact below.

Thus, in spite of its practical attractiveness and its implication for constraining physical models and for risk assessment, we believe that M_{max} cannot be used fruitfully in applications. The purpose of the present chapter is to re-examine the issue of the tail of the earthquake magnitude distribution and to suggest other characteristics of the tail behavior that are more stable than M_{max}.

The problem of statistical characterization of extreme, rare events is reduced to estimation of quantiles of high level (close to unity), the estimate being based on a finite sample. We recall that the quantile Q_q of level q, $0 < q < 1$ of a continuous, monotone distribution function $F(x)$ is defined as the root of the equation

$$F(x) = q.$$

The problem of statistical estimation of quantiles of high level is extremely important for practice. Many applied problems boil down, in fact, to the estimation of such quantiles. As an example, we can mention the catastrophic collapse of dykes at several locations in the Netherlands (1953) during a severe storm that caused a disastrous flooding of large areas and killed over 1,800 people. An extreme event has occurred (a record surge) that has broken the sea dykes. In the wake of this catastrophe, the Dutch government established the Delta Committee [Ha]. The task of the committee was to answer the following question: "What would the dyke heights have to be (at various locations) so as to prevent future catastrophic flooding with a sufficiently high probability, e.g., 0.9999?" From the statistical standpoint, the problem was reduced to estimation of a quantile of level 0.9999 for maximum annual surge heights. What were available historical data? The basic data consisted of observations of maximum annual heights. The sea-water level in the Netherlands is typically measured in (N.A.P.+x) meters (N.A.P. stands for Normal Amsterdam Peil, the Dutch reference level corresponding to mean see level). The 1953 flood was caused by a (N.A.P.+3.85) m surge, whereas historical accounts estimate a (N.A.P.+4) m surge for the 1,570 flood, the worst recorded. The statistical analysis in the Delta Committee report led to an estimated $(1 - 10^{-4})$ quantile of (N.A.P.+5.14) m for the yearly maximum. That is, the one-ten-thousand-year surge height is estimated as (N.A.P.+5.14) m. We refer for details to the paper by de Haan [Ha]. We wish to remark on one important point in this investigation:

The task of estimating such a 10,000-year occurrence of an event leads to *estimation well beyond the range of the data.* We discuss below extrapolation methods of this type based on some general statistical limit theorems.

If $x_1 < x_2 < \cdots < x_n$ is an ordered sample of iid random variables, then the quantile Q_q of a *fixed* level q can be estimated by the sample quantile x_{k+1} (k = entire part of $q{\cdot}n$) as $n \to \infty$. This estimate is known to be consistent. However, it becomes questionable for quantiles of higher levels, when $1 - q \propto const/n$, in particular, when $q > 1 - 1/n$. But it is precisely such levels that are of the most practical interest. For example, the above example features a sample size equal to $n \cong 500$ (five centuries of annual maximum sea levels), so $1 - 1/n = 0.998$, whereas the committee needed the quantile of level $q = 0.9999$. The estimation of quantiles that are "out of sample range", i.e., for $q > 1 - 1/n$, can be provided only under some extra assumption about the distribution in question. There is no magical technique which yields reliable results for free. Rephrasing a financial truth one can say:

"There is no free lunch, when it comes to high quantile estimation!" We shall use for this purpose the assumption of validity for two main limit theorems in the extreme value theory (see, e.g., [EKM]). The first Theorem is formulated as follows.

Let us assume that $x_1,\ldots, x_n$ is a iid sequence of random variables with a continuous distribution function $F(x)$ and $M_n = max(x_1,\ldots, x_n)$.

Theorem 1. *If there are a_n , b_n such that $(M_n - a_n)/b_n$ converges weakly to some non-degenerate random variable with the distribution function $\Phi(x)$, then $\Phi(x)$ can be a function of only one of three types, apart from the location and scale parameters:*

$$\begin{array}{llll}
\Phi_1(x|\xi) = exp(-1/x^{1/\xi}), & x>0; & \xi>0; & \text{(Pareto type)} \\
\Phi_2(x|\xi) = exp(-|x|^{-1/\xi}), & x<0; & \xi<0; & \text{(Weibull type)} \\
\Phi_2(x|\xi) = 1; & x \geqslant 0; & & \\
\Phi_3(x) = exp(-exp(-x)), & |x|<\infty; & & \text{(Gumbel type)}
\end{array}$$

These three relations can be written in a unified form. If the parameters of location μ and scale σ are added, then the unified formula is written as follows:

$$\Phi(x|\mu,\sigma,\xi) = exp(-[1+\xi(x-\mu)/\sigma]^{-1/\xi}), \quad \xi \neq 0. \tag{6.2}$$

For a zero shape parameter this equation transforms into

$$\Phi(x|\mu,\sigma,0) = exp(-exp[-(x-\mu)/\sigma]), \quad \xi = 0; \tag{6.3}$$

The shape parameter ξ varies from minus to plus infinity. Its sign determines the domain of definition and the distribution type:

$$\begin{aligned} &\xi > 0, \quad \text{domain of definition}: \ 1+\xi(x-\mu)/\sigma \geqslant 0, \quad \text{Pareto type},\\ &\xi < 0, \quad \text{domain of definition}: \ 1+\xi(x-\mu)/\sigma \geqslant 0, \quad \text{Veibull type};\\ &\xi = 0, \quad \text{domain of definition}: |x| < \infty, \quad \text{Gumbel type}. \end{aligned}$$

The distribution (6.2)–(6.3) is called the Generalized Extreme Value distribution (GEV). We denote its distribution function as $GEV(x\,/\mu,\sigma,\xi)$:

$$GEV(x|\mu,\sigma,\xi) = \Phi(x|\mu,\sigma,\xi).$$

We see that for negative ξ there exists a finite rightmost boundary which for the case of earthquakes can be denoted as M_{max}:

$$M_{max} = \mu - \sigma/\xi. \tag{6.4}$$

We shall also use the Generalized Pareto Distribution (GPD):

$$G_h(x|\xi,s) = 1 - [1+\xi(x-h)/s]^{-1/\xi}. \tag{6.5}$$

The parameters (ξ, h, s) define the domain of definition:

$$\begin{aligned} &\text{if} \quad \xi \geqslant 0, \quad \text{then} \quad x \geqslant h;\\ &\text{if} \quad \xi < 0, \quad \text{then} \quad 0 \leqslant x \leqslant h - s/\xi. \end{aligned}$$

We see again that for negative ξ there exists a finite rightmost boundary M_{max}:

$$M_{max} = h - s/\xi. \tag{6.6}$$

Now the second main theorem of extreme value theory can be formulated as follows. We define the rightmost point x_F of the distribution function $F(x)$ as

$$x_F = sup\{x : F(x) < 1\};$$

x_F may be infinite. We define also the excess distribution function $F_h(x)$ as the conditional distribution

$$F_h(x) = P\{X - h < x|X > h\}.$$

Theorem 2. *If the condition of Theorem* 1 *is fulfilled and* $h \uparrow x_F$, *then there exists a positive normalizing function* $s(h)$ *depending on* h *such that*

$$sup|F_h(x) - G_h(x|\xi, s(h))| \to 0, \quad h \to x_F. \tag{6.7}$$

The supremum in (6.7) *is taken over* $0 \leq x \leq x_F - h$.

Thus, GPD is the limit for the excess distribution of any initial distribution function $F(x)$ that satisfies the requirement of a non-degenerate limit distribution existing for the normalized sample maximum. Note that the limit theorems in the extreme value theory are valid not only for series of iid random variables as sample size n tends to infinity, but also for samples of random size ν, under the condition that ν tends to infinity with probability one. We shall study maximum magnitudes on a time interval $(0, T)$. Assuming the sequence of main events to be a Poissonian stationary process with some intensity λ, we have the average number of observations on $(0, T)$ equal to λT. If $T \to \infty$, then the number of observations on $(0, T)$ tends to infinity with probability one and we can use (6.2), (6.5) as limit distributions of maximum main-shock magnitudes M_T on an expanding time interval $(0, T)$.

The limit distribution of maxima studied in the theory of extreme values can be derived in two ways. The first consists in directly selecting maxima of n successive observations $M_n = max(x_1,..., x_n)$, and studying their distribution as n tends to infinity. The second is to increase the threshold h of registered observations, and the method is sometimes called "peaks over threshold" method. In that case the distribution of maxima exceeding h tends (as h and n tend to infinity) to the Generalized Pareto Distribution (GPD), which is a function of two unknown parameters (ξ, s) and the known threshold h.

We give the following definition: the T-maximum of a stationary Poisson sequence of observations with the distribution function $F(x)$ is defined as the maximum observation for the time period T. If we denote the intensity of occurrence by λ, then the distribution function of the T-maximum (given at least one event takes place during this time period) is as follows:

$$\Psi_T(x) = \{exp(-\lambda T \cdot [1 - F(x)]) - exp(-\lambda T)\}/\{1 - exp(-\lambda T)\} \tag{6.8}$$

$$\cong exp(-\lambda T[1 - F(x)]), \quad if \quad \lambda T \gg 1. \quad \text{(Lomnitz formula)} \tag{6.9}$$

Three Corollaries of the Two Limit Theorems:

1. *The T-maximum will have the GEV-distribution (apart from terms of order exp (–λT)) only when F(x) is a GPD-distribution:* $F(x) = G_h(x/\xi, s)$ *for some* h, ξ, s.

This corollary enables us to compare results obtained by the estimation methods based on the GPD and GEV distributions.

2. *If a random variable X has a GPD distribution* $G_h(x/\xi, s)$, *then the conditional distribution of X given* $X > H > h$ *is also a GPD distribution* $G_h(x/\xi, S)$, *where*

$$S = s + \xi(H - h). \tag{6.10}$$

Based on this corollary, the s-estimates for different thresholds h can be combined in a joint estimation procedure.

3. *If a random variable X has a GPD distribution $G_h(x/\xi, s)$ and the maximum $M_T = max(X_1,..., X_\nu)$ is used, where ν is a Poisson random variable with the parameter λT (λT » 1), then M_T has a GEV distribution GEV(x / ξ,μ(T),σ(T)), apart from terms of the order exp(–λT), where*

$$\mu(T) = h - (s/\xi) \times [1 - (\lambda T)^{\xi}]; \tag{6.11}$$

$$\sigma(T) = s \times (\lambda T)^{\xi}. \tag{6.12}$$

The converse is true as well: if $M_T = max(X_1,..., X_\nu)$ follows a GEV-distribution (6.2), (6.3) *with parameters ξ, σ, μ, then the original distribution of X_k is a GPD-distribution* (6.5) *with the parameters*

$$s = \sigma \cdot (\lambda T)^{-\xi}; \tag{6.13}$$

$$h = \mu + (\sigma/\xi) \cdot [(\lambda T)^{-\xi} - 1]. \tag{6.14}$$

Again, we disregard here terms of order exp(–λT).

Corollary 3 is used below to convert the parameters of the GEV distribution into those for the corresponding GPD distribution. Proofs of these Corollaries are given in [PSSR]. We see that the shape parameters of GPD and GEV are always identical, whereas the location and scale parameters differ.

Using relations (6.11), (6.12), (6.13), and (6.14) one can convert estimates μ(T), σ(T) obtained for some T into corresponding estimates for another time interval τ:

$$\mu(\tau) = \mu(T) + (\sigma(T)/\xi) \cdot [(\tau/T)^{\xi} - 1]; \tag{6.15}$$

$$\sigma(\tau) = \sigma(T) \cdot (\tau/T)^{\xi}. \tag{6.16}$$

Relations (6.11), (6.12), (6.13), (6.14), (6.15), and (6.16) are very convenient and we shall use them in our estimation procedures. Here, the interval T belongs to the past time in the catalog (or part of the catalog) used for estimation, whereas τ refers to a future time interval (prediction).

From the GPD-distribution (6.5) or GEV-distribution (6.2), (6.3) we can get the quantiles $Q_q(\tau)$ that we suggest as stable, robust characteristics of the tail distribution of magnitudes. These quantiles are the roots of

$$GEV(x|\xi, \sigma, \mu) = q; \tag{6.17}$$

$$GPD_h(x|\xi, s) = q; \tag{6.18}$$

Inverting Eqs. 6.17 and 6.18, we get, respectively,

$$Q_q(\tau) = \mu(T) + (\sigma(T)/\xi) \cdot [a \cdot (\tau/T)^{\xi} - 1]; \tag{6.19}$$

$$Q_q(\tau) = h + (s/\xi) \cdot [a \cdot (\lambda\tau)^{\xi} - 1], \tag{6.20}$$

where $a = [log(1/q)]^{-\xi}$.

Thus, instead of the instable parameter M_{max} we suggest a more appropriate characteristic $M_{max}(\tau)$, which is the maximum earthquake in a future time interval τ. The random variable $M_{max}(\tau)$ can be described by its distribution function or by its quantiles $Q_q(\tau)$ that are, in contrast to M_{max}, stable and robust characteristics. Besides, if $\tau \to \infty$, then one has with probability one that $M_{max}(\tau) \to M_{max}$ (of course, if $\tau \to \infty$, then $Q_q(\tau)$ gradually loses its stability). The methods of calculation of $Q_q(\tau)$ are explained and illustrated below. In particular, we can estimate $Q_q(\tau)$ for, say, $q = 10\%$, 5% and 1%, or whatever is required, and the median ($q = 50\%$) for any desirable time interval τ.

6.2 Estimation of Quantiles $Q_q(\tau)$ with the Help of Theorem 1 (Fitting the GEV Distribution)

The main difficulty in parameter estimation with the help of Theorem 1 consists in the choice of a proper interval T for measuring T-maxima in the catalog. There are some restrictions on this choice, and we are going to discuss these in detail. Sometimes they lead to a unique admissible value of T. In that case we shall use Theorem 1 with this T. But often some interval $T_1 < T < T_2$ of admissible values remains that satisfies all restrictions. The shape parameter ξ does not depend on T, so the estimates of ξ corresponding to different T can be averaged, but the parameters μ and σ do depend on T (see (6.11), (6.12)). It follows that a simple averaging of their estimates is not proper. We need to find a way of a joint treatment to combine the estimates of these two parameters corresponding to different values of T. This can be done with the help of Eqs. 6.15, 6.16.

As noted above, the choice of a T-interval should satisfy the following contradictory GEV-restrictions:

1. The T-interval should be large enough, so that the limiting Theorem 1 can be used. In our case this demand is reduced to the condition that the mean number of events on each T-interval, λT, should be large enough. There should not be empty intervals (with no event). The Kolmogorov distances $KD(T)$ between the fitted GEV and the actual sample DF should be small enough. The estimation of significance level of KD should take into account the fact that the three fitted GEV-parameters decrease the quantiles of the Kolmogorov distribution (see Appendix A and [PSSR]). After some numerical experiments we found that the validity of the conditions for Theorem 1 requires $\lambda T \geq 7$, which provides the inequality $exp(-\lambda T) < 0.001$. In the case of discrete magnitudes, the quantiles of the Kolmogorov test are to be replaced with the quantiles of the corresponding chi-square test (see for details below).

2. The T-interval should be small enough, so that the sample size n of T-maxima ($n = T_{cat}/T$; T_{cat} is the time interval covered by the catalog) would be large enough to provide a reliable estimation of the three unknown parameters the GEV distribution involves. Our numerical experiments showed that n should be not be less than $60 \div 70$.
3. If the parameter ξ is negative, then T-values selected for the estimation should provide in accordance with (6.12), (6.11) a monotone decreasing estimate of $\sigma(T)$ and a monotone increasing estimate of $\mu(T)$.
4. The scatter of estimates should be taken into account when one selects T-values. The values of T leading to lower scatter should be preferred.

If these restrictions do not lead to a unique optimal value of T, but lead only to an interval of admissible values ($T_1 < T < T_2$), then it is necessary to choose within this interval a set of values $T_1 < T_2 < \cdots < T_r$ and for each T_k to derive statistical estimates of the GEV parameters:

$$\xi(T_1), \xi(T_2), \ldots, \xi(T_r); \tag{6.21}$$

$$\sigma(T_1), \sigma(T_2), \ldots, \sigma(T_r); \tag{6.22}$$

$$\mu(T_1), \mu(T_2), \ldots, \mu(T_r). \tag{6.23}$$

For the parameter ξ we average estimates (6.21), since their theoretical analogues do not depend on T. For the parameters μ, σ we select some reference time interval, say, $\tau = 10$ years (3,650 days), and convert (6.22), (6.23) into estimates of these parameters corresponding to τ using Eqs. 6.15 and 6.16. Then it is possible to average the estimates. Now, using Eq. 6.19, one can obtain the estimate of quantile $Q_q(\tau)$ for the averaged parameters:

$$Q_q(\tau) = \mu(\tau) + (\sigma(\tau)/\xi) \cdot [a \cdot (\tau/T)^{\xi} - 1]; \quad a = [log(1/q)]^{-\xi}. \tag{6.24}$$

6.3 Estimation of Quantiles $Q_q(\tau)$ with the Help of Theorem 2 (Fitting the GPD Distribution)

The main difficulty in parameter estimation with the help of Theorem 2 consists in the choice of a proper threshold h. There are some restrictions too for this choice. Sometimes they lead to a unique admissible value of h. In this case we shall use Theorem 2 with this value of h. But often some interval $h_1 < h < h_2$ of admissible values remains that satisfies all restrictions. The shape parameter ξ does not depend on h, so that its estimates for different thresholds can be averaged, but the parameter s does depend on h (see (6.10)). It follows that a simple averaging of estimates of this parameter is not proper. It is necessary to find some joint treatment of estimates of this parameter corresponding to different values of h. This can be done with the help of Eq. 6.10.

As we noted above, the choice of threshold h should satisfy the following contradictory GPD-restrictions similar to the GEV-restrictions:

1. The threshold h should be high enough, so that the limiting Theorem 2 can be used. The Kolmogorov distance *KD* between a fitted GPD and the actual sample DF should be small enough. The estimation of significance level for *KD* should take into account the fact that the two fitted GPD-parameters decrease the quantiles of the Kolmogorov distribution (see Appendix A and [PSSR]. In the case of discrete magnitudes, instead of quantiles of the Kolmogorov test it is common practice to use the quantiles of the corresponding chi-square test (see for details below).
2. Sample size n of observations exceeding h should be large enough to provide reliable estimates of ξ, s. Our numerical experiments showed that it is necessary to have $n > 70 \div 80$.
3. If ξ is negative, then the thresholds selected for the estimation should provide, in accordance with (6.10), monotone decreasing estimates $s(H)$.
4. The scatter of estimates should be taken into account when one selects a threshold. The values of h leading to lower bootstrap scatter should be preferred.

The shape parameter ξ does not depend on h (in the limit as $h \to \infty$), so the estimates of ξ corresponding to different h can be averaged, but the parameter s does depend on h (see Eq. 6.10), hence simple averaging of its estimates is not proper. It is necessary to find some joint treatment of estimates of this parameter corresponding to different values of h. This can be done with the help of Eq. 6.10.

If these restrictions do not lead to a unique optimal value of h, but only to an interval of admissible values ($h_1 < h < h_2$), then it is necessary to choose within this interval a set of values $h_1 < h_2 < \cdots < h_r$ and for each h_k to derive statistical estimates of ξ, s:

$$\xi(h_1), \xi(h_2), \ldots, \xi(h_r); \tag{6.25}$$

$$s(h_1), s(h_2), \ldots, s(h_r); \tag{6.26}$$

For the parameter ξ we average estimates (6.25). For parameter s we select some reference value of s, say, $s_1 = s(h_1)$, and convert estimates (6.26) into estimates of s_1 using Eq. 6.10. It is then possible to average them. Now, using Eq. 6.20, one can estimate the quantile $Q_q(\tau)$ for the averaged parameters:

$$Q_q(\tau) = h + (s/\xi) \cdot [a \cdot (\lambda\tau)^{\xi} - 1], \quad a = [log(1/q)]^{-\xi}. \tag{6.27}$$

Thus, we have two variants of equations for quantiles (6.24) and (6.27). As the methods for obtaining the estimates are different (although they use the same data), the quantiles will differ too. We can compare the results obtained using the GEV and GPD methods and in the case of their similarity draw a conclusion about the robustness of the estimate for the tail of the distribution of M_τ based on a given catalog. If the resulting quantiles differ significantly, then the estimation of tails is an ill-posed problem or perhaps, the data are insufficient.

We would like to emphasize once more that $Q_q(\tau)$ is a stable characteristic of the tail of the magnitude distribution, whereas the M_{max} parameter can be unstable. This can be shown as follows:

$$M_{max} = h - s/\xi.$$

If $\xi \to 0$, then $M_{max} \to \infty$, but $Q_q(\tau)$ remains finite if $q < 1$ is fixed.

$$Q_q(\tau) = h + (s/\xi) \cdot [a \cdot (\lambda\tau)^{\xi} - 1] \to h + s \cdot log[\lambda\tau/log(1/q)] < \infty. \qquad (6.28)$$

This result can be interpreted as "a mathematical proof" of the fact that the estimation of M_{max} can be intrinsically very unstable (at least, in regions of high seismicity where the parameter ξ can be near to zero), whereas the quantile $Q_q(\tau)$ is always stable.

The methods we have set forth above, concerning parameter estimation for the GEV and GPD distributions can be improved slightly with the bootstrap procedures [PSSR]. Specifically, before the estimation of the GPD parameters one can use a bootstrap procedure, that is, to generate n_b catalogs from the original catalog by random sampling with replacement. Having n_b samples (catalogs), one can estimate the parameters for each catalog, and then average the results. This procedure is supposed to somewhat diminish the random scatter of the estimates. Note that this averaging is applied to the fixed original catalog. Consequently, when n_b has been increased to reach some value, the decrease of the scatter is retarded. This is natural: two catalogs relevant to different fixed periods of time yield different estimates, even if $n_b \to \infty$; that is, the difference of averaged bootstrap estimates does not tend to zero. Our numerical experiments have shown that it is enough to take $n_b \cong$ 50–100, and then further increase of n_b hardly diminishes the variance. The situation in this case is quite analogous to that with the classical bootstrap.

A procedure called "shuffling" is analogous to the bootstrap. It can be used for GEV fitting. For this procedure, n_S catalogs should be generated. Each catalog has the same set of magnitudes as the actual catalog, but the time of each event is sampled at random from a uniform distribution over the whole time period covered by the original (actual) catalog. All generated catalogs have the same intensity λ (mean number of events per unit time) as the original one. Thus, the distribution of T-maxima M_T in shuffled catalogs will be the same as in the original catalog. Here we use one well-known property of the Poisson process, which is as follows: for a fixed number of Poisson points in an interval, their distribution on the interval is uniform. The next step is to estimate the GEV parameters for each of the n_S catalogs and to average the results. As above, there is no need to generate many catalogs: it is enough to take $n_S \cong$ 50–100.

The two procedures described above (the bootstrap and the shuffling) decrease the standard deviations of the estimates of parameters by factors of 1.1–1.4. At the same time, the estimates of the standard deviation based on the generated catalogs (to be called bootstrap-std or shuffling-std) remain unfortunately much lower than the respective true standard deviations. The true standard deviations can be estimated correctly only by independent extra data, or by the simulation procedure described below.

6.4 Application of the GEV and GPD to the Estimation of Quantiles $Q_q(\tau)$. The Global Harvard Catalog of Scalar Seismic Moments

We use the two methods described above for the Harvard catalog of seismic moments for the period from January 1, 1977 to June 16, 2006. We restrict the depth of sources to $h \leq 70$ km and magnitudes to $m_W \geq 5.5$ (or for seismic moments $\geq 2.24 \times 1e + 24$ dyne-cm). Note that this time interval contains the maximum earthquake, that of December 26, 2004, $m_W = 9.0$ (Sumatra). To eliminate aftershocks from the catalog, the space-time window suggested in [KK] was used:

$$(t;\ t + 10^{-0.31+0.46\,m}); \tag{6.29}$$

$$R_{km}(\varphi, \lambda; \varphi_0, \lambda_0) \leqslant 10^{-0.85+0.46\,m}, \tag{6.30}$$

where t, φ_0, λ_0, m are respectively the time, longitude, latitude and magnitude of the main shock. Time is measured in days and distances in kilometers. Magnitude m was converted from scalar seismic moment M_S by the relation

$$m = (2/3) \cdot (lg_{10}(M_S) - 16.1).$$

$R_{km}(\varphi_0,\lambda_0;\ \varphi,\lambda)$ is the distance between the main shock with coordinates φ_0,λ_0 and the current event under test with coordinates φ,λ. All shocks that fall in the window (6.29), (6.30) are regarded as aftershocks of the main shock and removed. The testing starts from the largest event and terminates when all eliminations cease. The sequence left after this testing is regarded as a Poisson sequence with intensity $\lambda(h) = \lambda$ as given by the number of events (which depends on a variable threshold) divided by the period of time covered by the catalog. The Poisson properties of such a sequence were confirmed by several statistical tests (see for details [PSSR]). The number of main shocks that have been left on aftershock elimination was $n = 4193$.

In order to test the Harvard catalog for stationarity in the time-magnitude window considered we plotted in Fig. 6.1 yearly intensities for the thresholds $h =$ 5.5; 5.7; 5.9 (main events). We see that for all three thresholds the time series of intensities can be considered as fairly stationary.

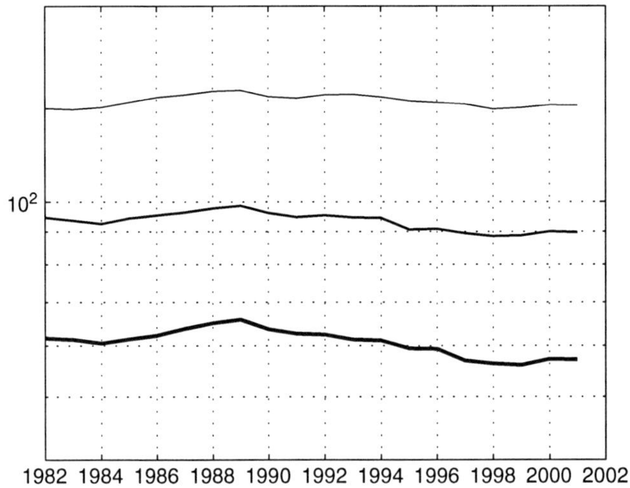

Fig. 6.1 The intensities (1/year) for the Harvard catalog. Thresholds: $h = 5.5$ (*top*); $h = 5.7$ (*middle*); $h = 5.9$ (*bottom*)

The GEV parameters were estimated by the method of moments, which is more effective here than the maximum likelihood method for samples of small and moderate size (for further details see [PSSR]).

As noted above, the choice of the T-interval should satisfy the GEV-restrictions listed in the preceding section.

The number N_T of different T-intervals in the catalog and corresponding products λT are given below:

T	50	75	100	125	150	175	200	225	250
N_T	214	143	107	85	71	61	53	47	42
λT	19.5	29.2	39.0	48.7	58.5	68.2	77.9	87.7	97.4

The intensity λ for $h = 5.5$ is determined as the ratio (*number of events* ≥ 5.5)/(*number of days*) $= 4{,}193/10{,}759 = 0.3897$. None of the T-intervals was empty for $T > 20$. The first GEV-restriction $\lambda T > 7$ becomes $T > 20$ in our case.

Figure 6.2 shows *KD*-distances for $20 \leq T \leq 250$. In order to estimate their significance levels we used the method proposed in [St] for the evaluation of significance levels in *KD*-testing with fitted parameters. 1,000 random catalogs were generated obeying a GEV-distribution with parameters equal to some typical values; these were taken to be equal to the estimates based on the Harvard catalog with $T = 100$, namely:

$$\xi = -0.190; \quad \mu = 7.245; \quad \sigma = 0.439. \tag{6.31}$$

We estimated for each synthetic catalog its GEV-parameters, inserted these estimates into the distribution function (6.2), and calculated the standard *KD*

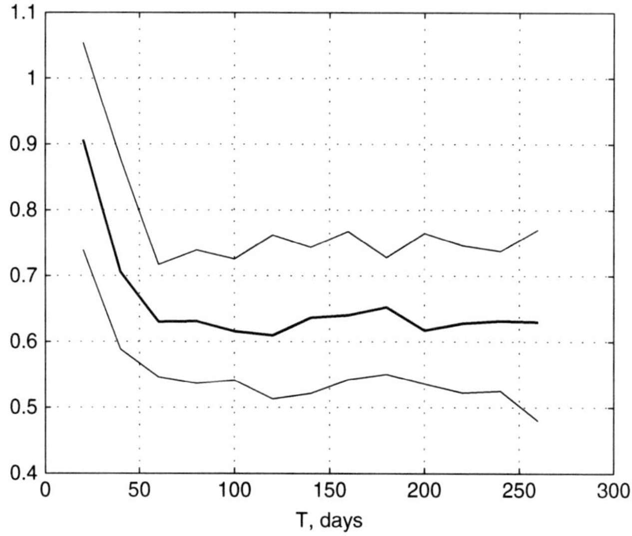

Fig. 6.2 Kolmogorov distances (*thick line*) between sample DF and GEV-distribution (the Harvard catalog, 100 shufflings) with 16% and 84% sample quantiles (*thin lines*)

distance (Appendix A). Then we were able to estimate the significance level ω of observed *KD*(*T*):

$$\omega = m/1{,}000; \quad m = \text{number of catalogs, where } KD \text{ was more observed } KD(T).$$

We got $\omega \cong 30\%$ for *KD*(*T*), corresponding to $50 < T < 180$ with $KD(T) \cong 0.61 \div 0.63$. These estimates enable us to say that the *KD*-test does not contradict the validity of the GEV-distribution for $50 < T < 180$. For $KD(T) = 0.9;\ 0.8$ the significance levels are, respectively, 1.6%; 4.8%. Thus, the values $T < 50$ should be rejected.

The *T*-interval should be large enough, so that the sample size n of *T*-maxima ($n = T_{cat}/T$; T_{cat} is the time interval covered by the catalog) would be large enough to provide reliable estimation of the three unknown parameters. The values $T > 250$ should not be used, since in this case the number of *T*-intervals ($n < 42$) is not sufficient for reliable estimation of the three parameters in the GEV distribution. Thus, having taken into account the GEV-restrictions (1, 2), we confine the interval of *T*-values to within (50; 250).

The estimates of the GEV-parameters ξ, μ, σ obtained by the method of statistical moments (see Appendix B) are shown in Fig. 6.3a, b, and c. In order to diminish the scatter of the estimates $n_s = 100$, shufflings were used for each value of T. We see some stabilization of the shape estimates in the range $100 \leq T \leq 150$. Figure 6.3b shows that the μ-estimates begin to increase only from $T \geq 100$ onward. In accordance with GEV-restriction (3), we have to reject estimates corresponding

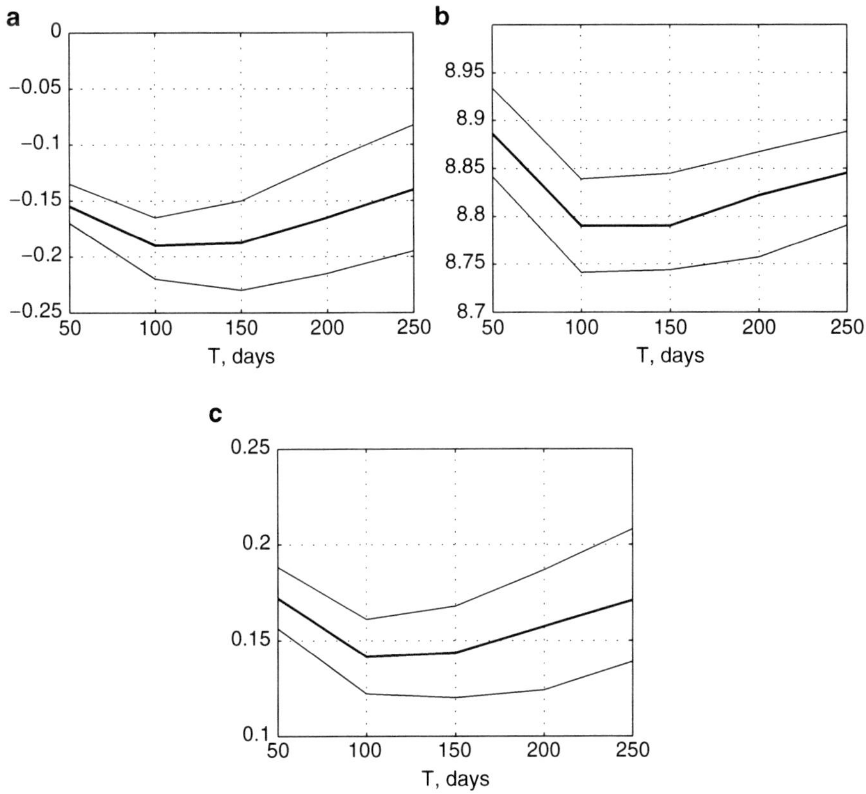

Fig. 6.3 (**a**) The ξ-estimates for different time intervals T (days). The Harvard catalog, 100 shufflings. (**b**) The μ-estimates for different time intervals T (days). The Harvard catalog, 100 shufflings. (**c**). The σ-estimates for different time intervals T (days). The Harvard catalog, 100 shufflings. Thin lines show 16% and 84% sample quantiles

to $T < 100$. Figure 6.3c shows that σ-estimates start to grow for $T > 150$. Again, using GEV-restriction (3), we can disregard $T > 150$. Taking into account these restrictions, we thus have the two values $T = 100$, 150 left for ultimate estimation. By the way, they give very similar estimates of our parameters. We have thus averaged the medians of shuffled estimates corresponding to $T = 100$ and 150 and got the final estimate of ξ:

$$\hat{\xi} = (-0.190 - 0.189)/2 \cong -0.190. \tag{6.32}$$

As to estimates of μ, σ, we recall that their theoretical values depend on T (see (6.11), (6.12)). Thus, in order to average these, one has to reduce them to comparable form. We used to this end Eqs. 6.15 and 6.16 and the standard time interval $\tau = 10$

years, i.e., 3,650 days, which we shall use below for the comparison of estimates related to different values of T. We got

$$\hat{\mu}_{10} = (8.383 + 8.381)/2 = 8.382; \tag{6.33}$$

$$\hat{\sigma}_{10} = (0.220 + 0.219)/2 \cong 0.220 \tag{6.34}$$

Now we can derive estimates of M_{max} and $Q_{0.95}(10)$ *by* inserting the mean parameters (6.32), (6.33), (6.34) into (6.19). We get

$$M_{max} = 9.54; \tag{6.35}$$

$$Q_{0.95}(10) = 8.88. \tag{6.36}$$

It should be noted that the 84% and 16% quantiles of the shuffled samples shown in Figs. 6.2 and 6.3 characterize the scatter of estimates for a fixed given catalog, and are thus below the true scatter derived from a set of independent catalogs, as explained at the end of Section 6.3. Thus, the 16% *and* 84% quantiles shown in these figures are conditional quantiles given a fixed catalog. In order to estimate the unconditional scatter, we have generated again 1,000 catalogs which were distributed according to the GEV with $T = 100$ and parameters (6.33), (6.34). Having used 1,000 synthetic catalogs, we estimated the standard deviation (std) for each parameter. The distribution of the estimates of M_{max} has a heavy tail (the consequence of instability of this parameter), and besides it is skewed. For this reason it is more appropriate to characterize its spread by the 16% and 84% quantiles (percentiles), $q_{0.16}$ and $q_{0.84}$. In the case of the Gaussian distribution the spread, i.e., difference $(q_{0.84} - q_{0.16})$, equals double the standard deviation. We assume that std-estimates (and quantiles) in our synthetic catalogs are close to the real ones. So, these std (and quantiles) were taken as statistical estimates of the std (spread) in the real catalog.

Thus, the following estimates by GEV-fitting have been obtained:

$$\hat{\xi} = -0.190 \pm 0.060; \tag{6.37}$$

$$\hat{\mu}_{10} = 8.382 \pm 0.094; \tag{6.38}$$

$$\hat{\sigma}_{10} = 0.220 \pm 0.039; \tag{6.39}$$

$$M_{max} = 9.54; \quad q_{0.16} = 8.99; \quad q_{0.84} = 10.29; \tag{6.40}$$

$$Q_{0.95}(10) = 8.88 \pm 0.210; \tag{6.41}$$

If we take for an estimate of the std of M_{max} its half-spread, we get

$$M_{max} = 9.54 \pm 0.65. \tag{6.42}$$

It should be noted that the std of M_{max} obtained as usual from shuffled catalogs is generally much larger than (6.42) obtained through the spread. It is a consequence of the heavy-tailed distribution of M_{max}-estimates. Comparing (6.41) and (6.42), we see that the std of M_{max} is approximately three times the std of $Q_{0.95}(10)$.

Now we are going to fit the GPD distribution to the Harvard catalog. The GPD parameters were estimated by the maximum likelihood method (see Appendix C).

We have the following sample sizes for different thresholds:

$$h \quad 6.0 \quad 6.2 \quad 6.4 \quad 6.6 \quad 6.8 \quad 7.0 \quad 7.2 \quad 7.4 \quad 7.6 \tag{6.43}$$

$$n_k \quad 1454 \quad 941 \quad 625 \quad 392 \quad 264 \quad 174 \quad 114 \quad 80 \quad 50 \tag{6.44}$$

Thresholds above 7.4 are hardly ever applicable, since in this case the number of large events exceeding the threshold is 80 or less. Such sample sizes are insufficient for reliable estimation of two parameters of the GPD distribution. We thus adopt the restriction $h \leq 7.2$.

Figure 6.4 shows *KD*-distances for $6.4 \leq h \leq 7.2$. Their minimum is close to 0.9 ($h = 6.8$), whereas $KD \cong 1.3$ for $h = 6.4$. In order to estimate its significance level we used the same method as earlier in the GEV fitting. 1,000 random catalogs were generated obeying the GPD-distribution with parameters equal to those obtained for the Harvard catalog by maximum likelihood and with the threshold $h = 6.8$, namely: $\xi = -0.197$; $s = 0.542$; $h = 6.8$. We estimated for each catalog the parameters ξ, s, inserted these estimates into the distribution function (6.5), and

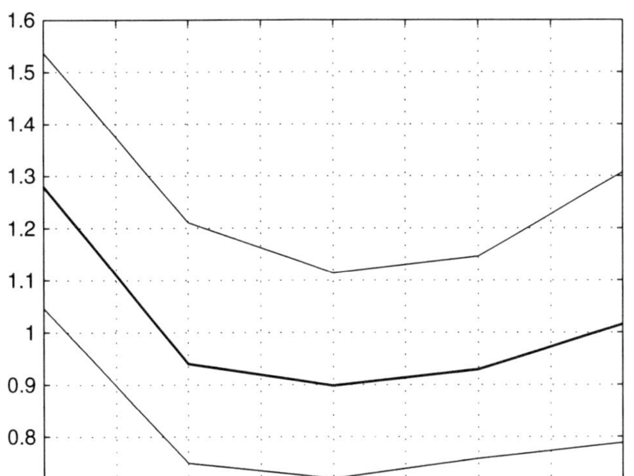

Fig. 6.4 Kolmogorov distances (*thick line*) between sample DF and GPD-distribution (the Harvard catalog, 100 bootstraps) with 16% and 84% sample quantiles (*thin lines*)

calculated the standard *KD* distance. Then we were able to estimate the significance level ω of observed *KD*(h) based on these 1,000 experiments:

$$\omega = m/1{,}000,$$

where m *is* the number of catalogs with *KD* larger than the observed *KD*(h).

We got $\omega \cong 40\% \div 50\%$ for $6.6 \le h \le 7.0$ and $\omega = 1\%$ for $h = 6.4$. These estimates enable us to say that the *KD*-test does not contradict the validity of the GPD-distribution for observations exceeding the thresholds $h = 6.6$; 6.8; 7.0, and agrees with the threshold $h = 6.4$. The threshold $h = 7.2$ corresponds to the probability 13% of exceeding the observed *KD*. Figure 6.5 shows s-estimates. We see again that for $h = 6.4$ the GPD-restriction (3) is violated. Besides, the spread of ξ-estimates for $h = 7.2$ is very large, as can be seen in Fig. 6.6. Thus, summing up all these restrictions, we leave for further analysis the interval of thresholds $6.6 \le h \le 7.0$. Table 6.1 shows estimates of parameters for these thresholds.

We see that these estimates are not greatly different for different thresholds. In such a situation, simple averaging is justified. Thus, taking the mean, we get

$$<\xi> \; = -0.197; \tag{6.45}$$

$$<s> \; = 0.534. \tag{6.46}$$

Using these estimates and the mean threshold $h = 6.8$, we get (from (6.6)) estimates of M_{max}, $Q_{0.95}(10)$:

$$<M_{max}> \; = 9.51; \tag{6.47}$$

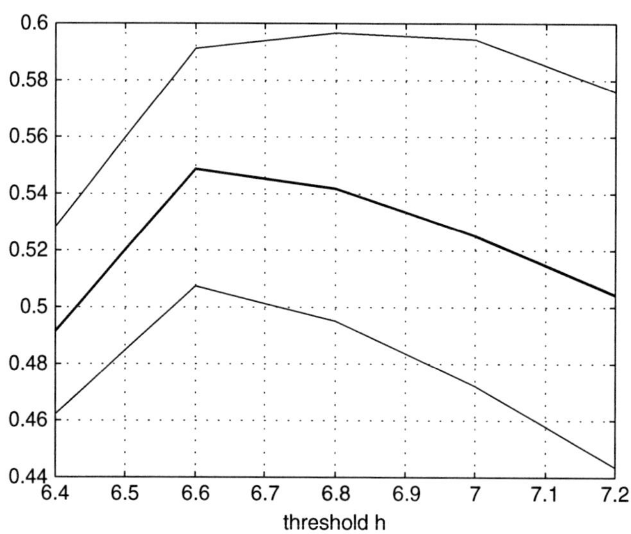

Fig. 6.5 The s-estimates (*thick line*) for different time intervals T (days). The Harvard catalog, 100 bootstraps with 16% and 84% sample quantiles (*thin lines*)

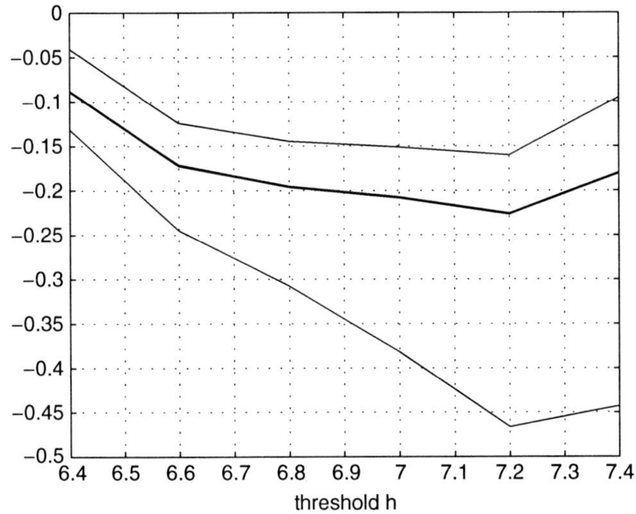

Fig. 6.6 The ξ-estimates (*thick line*) for different time intervals T (days). The Harvard catalog, 100 bootstraps with 16% and 84% sample quantiles (*thin lines*)

Table 6.1 Maximum likelihood estimates of parameters derived by GPD fitting

h	6.6	6.8	7.0
ξ	−0.173	−0.197	−0.214
s	0.549	0.542	0.525
M_{max}	9.77	9.56	9.47
$Q_{0.95}(10)$	8.95	8.91	8.91

M_{max} is the rightmost point of the magnitude distribution given by (6.6). $Q_{0.95}(10)$ is the 95% quantile of the maximum magnitude distribution (T-maximum magnitude) in 10-year intervals. $n_b = 100$ bootstraps were used for each threshold

$$< Q_{0.95}(10) > \; = 8.89. \tag{6.48}$$

In order to estimate the unconditional scatter, we have generated again 1,000 synthetic catalogs that were distributed according to the GPD with $h = 6.8$ and the parameters (6.44), (6.45). Having used the sample consisting of 1,000 such catalogs, we estimated the std for ξ, s, $Q_{0.95}(10)$ and 16%, 84% quantiles for M_{max}. Thus, the following estimates by GPD-fitting have been obtained:

$$\xi = -0.197 \pm 0.054; \tag{6.49}$$

$$s = 0.541 \pm 0.044; \tag{6.50}$$

$$M_{max} = 9.51; \quad q_{0.16} = 9.01; \quad q_{0.84} = 10.10; \tag{6.51}$$

$$Q_{0.95}(10) = 8.89 \pm 0.23. \tag{6.52}$$

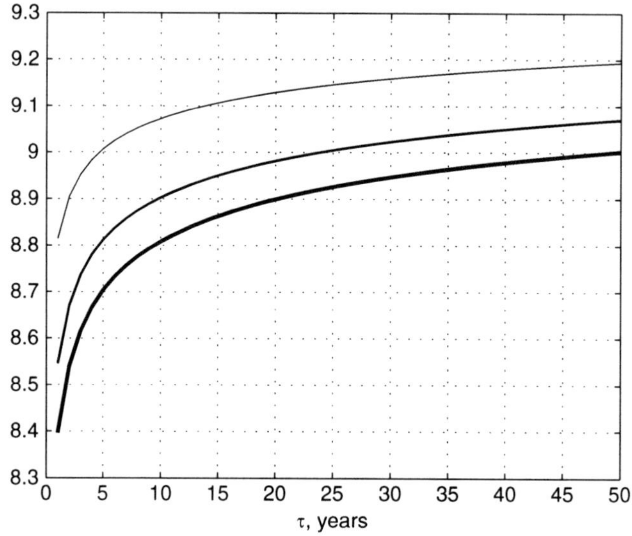

Fig. 6.7 99% (*top*), 95% (*middle*), 90% (*bottom*) quantiles of the maximum magnitude distribution for a future time interval τ (year)

If we take for an estimate of the std of M_{max} its half-spread, we get

$$M_{max} = 9.54 \pm 0.545. \tag{6.53}$$

Again, the spread of quantiles $Q_{0.99}(10)$ is substantially less than that of the M_{max}.

In order to illustrate the methods we have proposed, we give several plots of quantiles $Q_q(\tau)$ for a set of future intervals τ. Figure 6.7 shows quantiles $Q_q(\tau)$ for τ up to 50 years as obtained by fitting the GPD to the Harvard catalog. We used the parameter values $h = 6.8$; $\xi = -0.197$; $s = 0.541$ that were obtained in the GPD fitting. We see that the quantiles grow rather slowly, approaching the uppermost possible limit M_{max} somewhere around 9.5 (this figure should be considered as quite uncertain, as we stressed above). The quantiles in Fig. 6.7 characterize the uppermost tail of the distribution of T-maximum. In order to illustrate a "middle" tendency of the T-maximum distribution we show in Fig. 6.8 three quantiles with levels $q_1 = 0.16$; $q_2 = 0.50$ (*median*); $q_3 = 0.84$.

6.5 Application of the GEV and GPD to the Estimation of Quantiles $Q_q(\tau)$ for Catalogs of Binned Magnitudes

The considerations set forth in previous sections of this chapter deal with continuous random variables, with continuous distribution functions. For continuous random variables, the Kolmogorov test was used successfully in preceding sections.

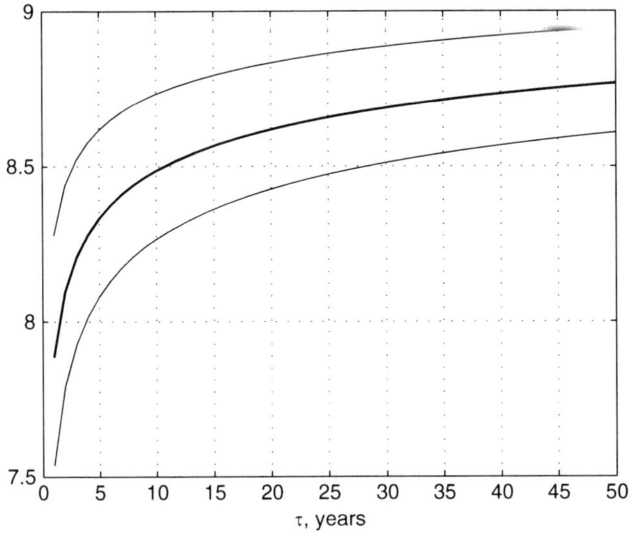

Fig. 6.8 84% (*top*), 50% (*middle*), 16% (*bottom*) quantiles of the maximum magnitude distribution for a future time interval τ (year)

For discrete rv the theory of extreme values is not directly applicable. But magnitude catalogs provide just discrete rv. We have to obviate this obstacle. Usually, in most existing catalogs, including the catalogs of Japan and Fennoscandia that we are going to consider, the bin width Δm equals 0.1. In some catalogs two digits after the decimal point are reported, but the last digit is fictitious, unless the magnitudes are calculated from seismic moments that have several exact digits (like seismic moment magnitude m_W in the Harvard catalog). We assume that digitization is fulfilled in intervals $((k-1)\cdot\Delta m;\ k\cdot\Delta m)$, k is an entire number. Then, in the GPD approach we should use half-integer thresholds $h = (k-1/2)\cdot\Delta m$, which is not a serious restriction. Further, having a sample of observations exceeding some $h = (k-1/2)\cdot\Delta m$, and fitting a GPD-distribution to it, we cannot use statistical tools tailored for continuous rv like the Kolmogorov test and the Anderson-Darling test. Such statistical tools are inapplicable to discrete variables. We calculated the Kolmogorov distances for digitized generated samples obeying the GEV-distribution and found that their distribution is very far from the true one (the Kolmogorov distances for digitized variables are much greater than those for continuous random variables, particularly for large samples). We are thus forced to use statistical tools devised for discrete rv. We have taken among such tools the standard χ^2-method (chi-square test) which provides both estimation of unknown parameters and a strict evaluation of goodness of fit.

The chi-square test has two peculiarities:

1. In order to be able to apply the chi-square test, a sufficient number of observations is needed in each bin (we choose this minimum number as being equal to 8

2. In studying the tail of a distribution it is desirable to make binning as detailed as possible at the extreme ends of the data range, since it is exactly the greatest observations which largely determine the properties of limit GEV or GPD distributions

In general, the chi-square test is less sensitive and less efficient than the Kolmogorov test or the Anderson-Darling test. This results from the fact that the chi-square coarsens data by putting them into discrete bins. It should be noted that, unlike continuous testing, the chi-square testing leaves a considerable uncertainty in its application, such as the choice of bin width, number of bins, and the initial point of the first bin. Unfortunately, this uncertainty cannot be completely removed.

Now we consider first the GPD fitting to the catalog of Japanese earthquakes.

6.5.1 Catalog of the Japan Meteorological Agency (JMA)

The full JMA catalog covers the area within 25.02 $\leq$ latitude $\leq$ 49.53° and 121.01 $\leq$ longitude $\leq$ 156.36° and the temporal window 01.01.1923–30.04.2007. The depths of earthquakes fall in the interval 0 $\leq$ depth $\leq$ 657 km. The magnitudes are expressed in 0.1-bins and vary in the interval 4.1 $\leq$ magnitude $\leq$ 8.2. There are 39,316 events in this space-time domain. The spatial domain covered by the JMA catalog includes the Kuril Islands and the east margin of Asia.

The focus of our study is on earthquakes occurring in the area of the central Japan islands. We thus restrict our study to the earthquakes occurring within the polygon with coordinates [(160.00; 45.00); (150.00; 50.00); (140.00; 50.00); (130.00; 45.00); (120.00; 35.00): (120.00; 30.00); (130.00; 25.00); (150.00; 25.00); (160.00; 45.00)]. Figure 6.9 shows a map of the area enclosed within the polygon. There were 32,324 events within this area. The corresponding frequency-magnitude relation is plotted in Fig. 6.10 and the histogram of magnitudes is shown in Fig. 6.11.

Next, we only retain shallow earthquakes whose depths are smaller than 70 km. We then applied the declustering Knopoff-Kagan space-time window algorithm with parameters (6.29), (6.30). The remaining events constitute our "main shocks" to which we are going to apply the GEV and the GPD methods described above. There are 6,497 main shocks in the polygon shown in Fig 6.9 with depths less than 70 km. The frequency-magnitude curve of these main shocks is shown in Fig. 6.12. It should be noted that the b-value (or slope) for the main shocks is significantly lower (by 0.15 or so) than the corresponding b-value for all events. From the relatively small number of remaining main shocks, one concludes that the percentage of aftershocks in Japan is very high (about 80% according to the Knopoff-Kagan algorithm). The histogram of these main events is shown in Fig. 6.13. One can observe irregularities and a non-monotonic behavior of the magnitude histogram. These irregularities force us to combine the 0.1 bins into 0.2 bins. This discreteness in magnitudes requires a special treatment (in particular, the use of the chi-square test), which is explained in the next subsection. On a positive note, no visible pattern associated with half-integer magnitude values can be detected.

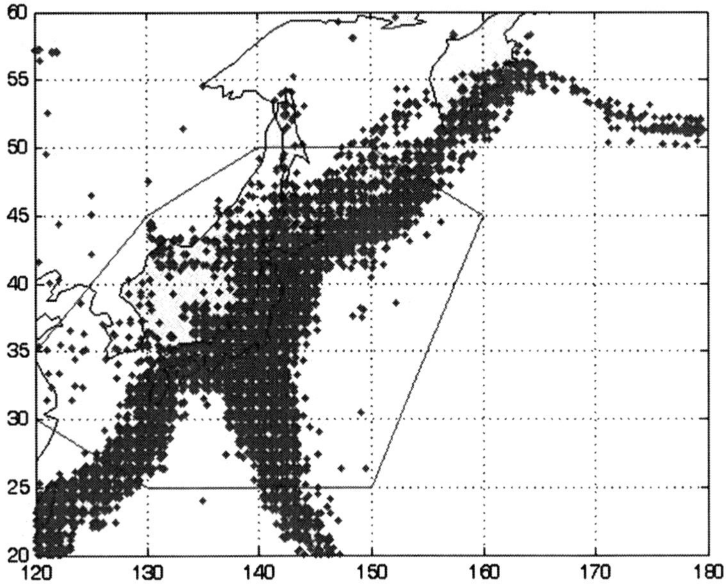

Fig. 6.9 The contour map of the Japan region with the polygon of study and earthquake epicenters, 1923–2007

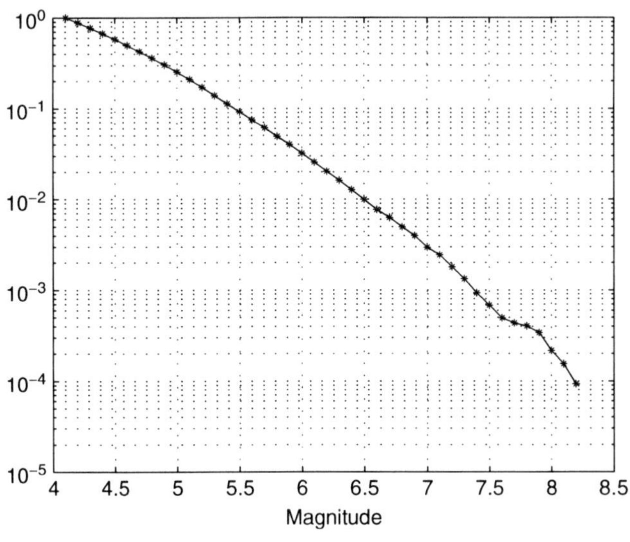

Fig. 6.10 The complementary sample distribution function $1 - F(x)$, Japan, all 1923–2007 events, $n = 32{,}324$

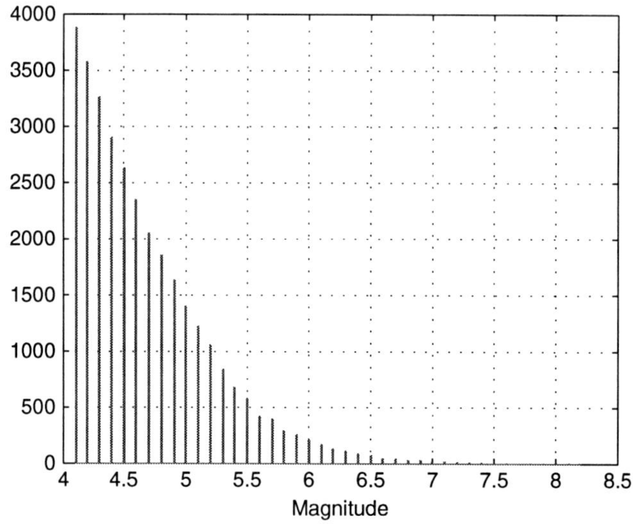

Fig. 6.11 The histogram of all events, Japan, 1923–2007, $n = 32{,}324$

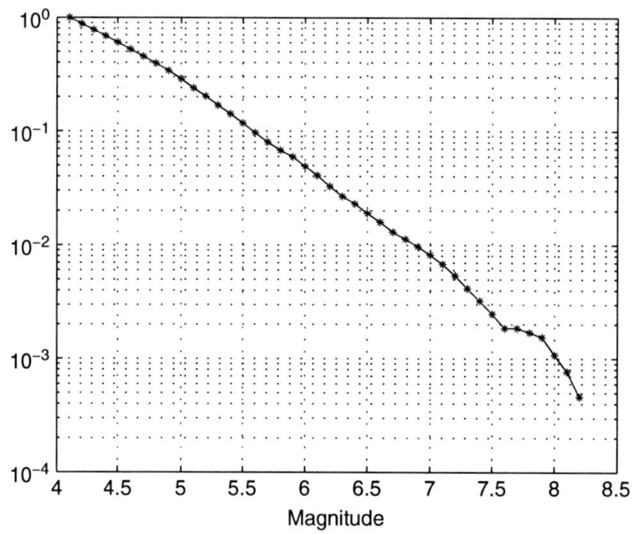

Fig. 6.12 The complementary sample distribution function $1 - F(x)$, Japan, 1923–2007 main shocks, $m \geq 4.0$; $n = 6{,}497$

Figure 6.14 shows a plot of yearly numbers of earthquakes (main events) averaged over 10 years for three magnitude thresholds: $m \geq 4.1$ (all available main events); $m \geq 5.5$; $m \geq 6.0$. For high magnitudes ($m \geq 6.0$), the sequence of events is approximately stationary.

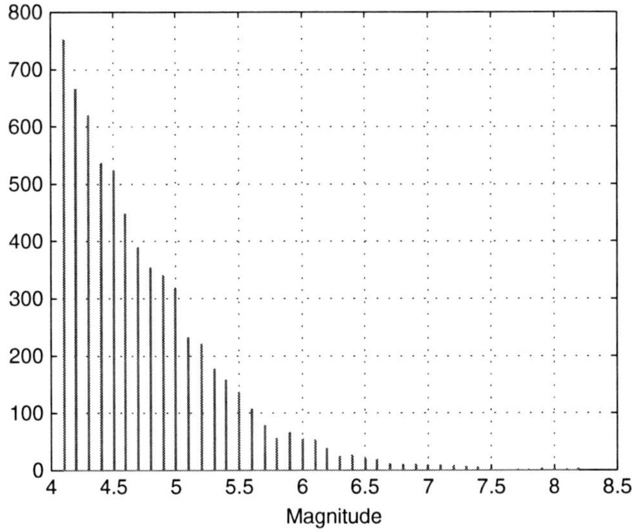

Fig. 6.13 The histogram of main shocks, Japan, 1923–2007, $m \geq 4.0$; $n = 6{,}497$

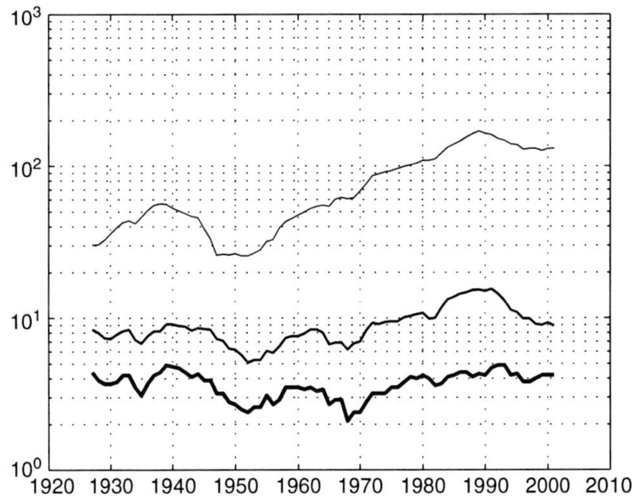

Fig. 6.14 The intensities (1/year) for the Japan catalog. Thresholds: $h = 4.0$ (*top*, $n = 6497$); $h = 5.7$ (*middle*, $n = 762$); $h = 5.9$ (*bottom*, $n = 319$)

6.5.1.1 The GPD Fitting

Consider the discrete set of magnitudes recorded at a step Δm over threshold h,

$$h + (k-1)\Delta m/2 \leqslant m < h + k\Delta m/2; \quad k = 1, \ldots r; \quad \Delta m = 0.1. \tag{6.54}$$

We assume that these "peaks over threshold", i.e., magnitudes exceeding the threshold h, obey a GPD distribution $G_h\ (x\ /\ \xi,s)$ with some unknown parameters. The corresponding discrete probabilities are

$$p_k(\xi,s|h) = P\{h + (k-1)\times 0.05 \leqslant m < h + k\times 0.05\}$$
$$= G_h(h + k\times 0.05|\xi,s) - G_h(h + (k-1)\times 0.05|\xi,s); \quad k = 1,\ldots,r \tag{6.55}$$

$$p_{r+1}(\xi,s|h) = P\{h + r\times 0.05 \leqslant m\} = 1 - G_h(h + r\times 0.05|\xi,s). \tag{6.56}$$

The last $(r+1)$-th bin covers the interval $(h + r\times 0.05; M_{max})$. Using (6.5), we get

$$G_h(x|\xi,s) = 1 - [1 + (\xi/s)\cdot(x-h)]^{-1/\xi}, \quad h \leqslant x \leqslant h - s/\xi, \quad \xi < 0.$$

Let us assume that the k-th bin contains n_k observations. Summing over the $r+1$ bins, the total number of observations is $n = n_1 + n_2 + \cdots + n_r + n_{r+1}$. The chi-square sum $S(\xi,\ s)$ is then written as follows:

$$S(\xi,s) = \sum_{k=1}^{r+1} [n_k - n\cdot p_k(\xi,s|h)]^2/n\cdot p_k(\xi,s|h), \tag{6.57}$$

$S(\xi,\ s)$ should be minimized over the parameters $(\xi,\ s)$. This minimum is distributed asymptotically as the chi-square with $(r-2)$ degrees of freedom. The quality of the fit of the empirical distribution by (6.55), (6.56) is quantified by the probability $P_{exc} = P\{\chi^2(r-2) \geq min(S)\}$, where $\chi^2(r-2)$ is the chi-square random variable with $(r-2)$ degrees of freedom, i.e., P_{exc} is the probability of exceeding the minimum fitted chi-square sum. The larger P_{exc}, the better is the goodness of fit.

For magnitude thresholds $h \leq 5.95$ and $h \geq 6.65$, the chi-square sums $min(S)$ turned out to be very large, leading to very small P_{exc} values, indicating that such thresholds are not acceptable. For thresholds in the interval $(6.05 \leq h \leq 6.55)$, the results of the chi-square fitting procedure are listed in Table 6.2. In order to obtain these results, we also have performed $n_b = 100$ bootstrapping procedures on our initial sample and took medians of the estimates, as described above.

In Table 6.2, three thresholds $h = 6.15$: $h = 6.25$ and $h = 6.35$ give very similar estimates. In contrast, the estimates obtained for the thresholds $h = 6.05$ and $h = 6.45$ have lower goodness of fit values (smaller P_{exc}). This suggests accepting the estimates that have the highest goodness of fit ($h = 6.25$). An alternative approach would consist in averaging the parameters for the three best thresholds, as was done in Section 6.4, but the resulting quantiles $Q_q(\tau)$ turned out to be very similar. So, we adopted a simpler approach taking the one threshold ($h = 6.25$) with the highest goodness of fit. We got

$$\xi_{GPD} = -0.214; \quad s_{GPD} = 0.640; \quad M_{max,\,GPD} = 9.31; \quad Q_{0.95}(10) = 8.29. \tag{6.58}$$

Table 6.2 Estimates provided by the chi-square fitting procedure using the GPD approach

h	6.05	6.15	6.25	6.35	6.45
r	7	7	6	6	6
Degrees of freedom	5	5	4	4	4
ξ	−0.0468	−0.2052	−0.2137	−0.2264	−0.1616
s	0.5503	0.6420	0.6397	0.6264	0.6081
M_{max}	17.87	9.43	9.31	9.11	10.20
$Q_{0.95}(10)$	8.73	8.32	8.29	8.24	8.52
P_{exc}	0.0753	0.2791	0.3447	0.3378	0.1747

The parameters are estimated by minimizing $S(\xi,s)$ as given by (6.52). M_{max} is the rightmost point of the magnitude distribution given by (6.6). $Q_{0.95}(10)$ is the 95% quantile of the maximum magnitude distribution (T-maximum magnitude) in 10-year intervals. 100 bootstraps were used

These estimates are very close to their mean values obtained for the three thresholds $h = 6.15$; 6.25; 6.35. In order to estimate the statistical scatter of these estimates, we simulated our whole procedure of estimation 1,000 times by generating GPD-samples with the parameters given by (6.53). We digitized the continuous data and applied the chi-square test to the binned observations. Combining the std estimates derived from simulations with the mean values (6.58) based on the actual catalog, the final results of the GPD approach for the JMA catalog can be summarized by

$$\xi_{GPD} = -0.214 \pm 0.103; \quad s_{GPD} = 0.640 \pm 0.063; \tag{6.59}$$

$$M_{max,GPD} = 9.31 \pm 1.14; \quad Q_{0.90}(10) = 8.29 \pm 0.49; \tag{6.60}$$

$$quantiles\ of\ M_{max}: \ q_{0.16} = 8.17; \quad q_{0.84} = 13.37. \tag{6.61}$$

One can observe that the std of M_{max} exceeds the std of $Q_{0.90}(10)$ by a factor larger than two, confirming once more our earlier conclusion on the instability of M_{max}.

6.5.1.2 The GEV Fitting

In this approach, we divide the total time interval T_{cat} from 1923 to 2007 covered by the catalog into a sequence of non-overlapping, contiguous intervals of length T. The maximum magnitude $M_{T,j}$ on each T-interval is measured. We have $k = [T_{cat}/T]$ T-intervals, so the sample of our T-maxima $M_{T,1}$,. . ., $M_{T,k}$ is of size k. The larger the value of T, the more accurate is the GEV-approximation on this interval, but one cannot choose too large a T, because the sample size k of the set of T-maxima would be too small. This would make inefficient the statistical estimation of the three unknown parameters (ξ, μ_T, σ_T). Besides, we should keep in mind the restrictions mentioned above which are imposed by the chi-square method, viz., that the minimum number of observations per bin should not be less than 8. In order to satisfy these contradictory constraints, we have compromised by restricting the T-values to be sampled in the rather small interval

$$200 \leqslant T \leqslant 300 \text{ days}. \tag{6.62}$$

It should be noted that, for all T-values over 50 days, the parameter estimates do not vary much, and it is only for $T \leq 40$ that the estimates are drastically different. We have chosen $T = 200$ and derived estimates minimizing the chi-square sum:

$$S(\xi,\mu,\sigma) = \sum_{j=1}^{k+1} [n_j - n \cdot p_j(\xi,\mu,\sigma)]^2 / n \cdot p_j(\xi,\mu,\sigma) \tag{6.63}$$

where

$$p_1(\xi,\mu,\sigma) = P\{m < h_0 + 1 \times 0.05\} = G_{GEV}(h_0 + 0.05|\xi,\mu,\sigma);$$

$$\begin{aligned} p_j(\xi,\mu,\sigma) &= P\{h_0 + (k-1) \times 0.05 \leqslant m < h_0 + k \times 0.05\} \\ &= G_{GEV}(h_0 + k \times 0.05|\xi,\mu,\sigma) \\ &\quad - G_{GEV}(h_0 - (k-1) \times 0.05|\xi,\mu,\sigma); \quad j = 2,\ldots,k; \end{aligned} \tag{6.64}$$

$$p_{k+1}(\xi,\mu,\sigma) = P\{m > h_0 + k \times 0.05\} = 1 - G_{GEV}(h + k \times 0.05|\xi,\mu,\sigma)).$$

Here, h_0 is some threshold chosen so that the conditions required for application of the chi-square test would be satisfied. In our case the threshold $h_0 = 5.95$ provides more than eight observations per bin with probability close to unity. The following estimates minimize the sum (6.63):

$$\xi_{GEV} = -0.190 \pm 0.072; \quad \mu_{GEV} = 6.339 \pm 0.038; \quad \sigma_{GEV} = 0.600 \pm 0.022; \tag{6.65}$$

$$M_{max,\,GEV} = 9.57 \pm 0.86; \quad Q_{0.90,\,GEV}(10) = 8.34 \pm 0.32. \tag{6.66}$$

$$\textit{quantiles of } M_{max} : q_{0.16} = 9.15; \quad q_{0.84} = 11.02. \tag{6.67}$$

The estimates of the scatter in (6.65), (6.66) were obtained by simulation with 1,000 samples, similarly to the method used in the GPD approach. In estimating the parameters we used the shuffling procedure described in Section 6.3 with $n_s = 100$ shufflings.

Finally, we show in Figs. 6.15 and 6.16 the quantiles $Q_q(10)$, $q = 0.90$; 0.95; 0.99, as functions of τ for $\tau = 1 \div 50$ years both for GPD and for GEV fitting. One can observe that the quantiles obtained by the two methods are very similar. This confirms the stability of the estimation. Figure 6.17 shows a plot of the GEV-median together with two accompanying quantiles (5% and 95%) that enclose the 90% confidence interval.

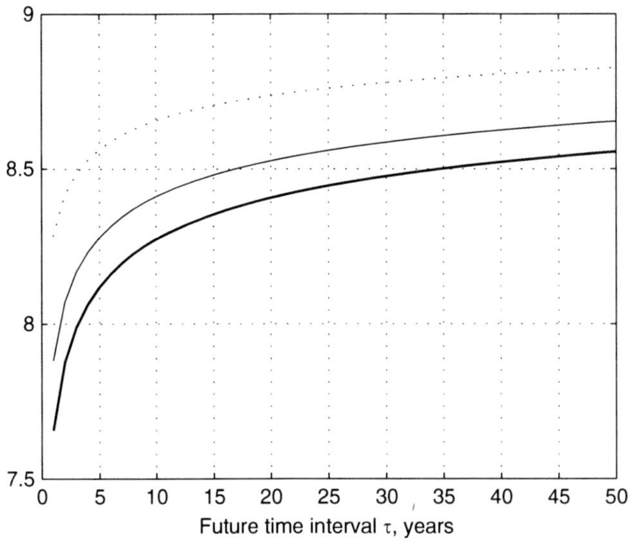

Fig. 6.15 The GPD-quantiles of the future τ-maxima, Japan. 99% (*top, dotted line*), 95% (*middle, light line*), 90% (*bottom, heavy line*)

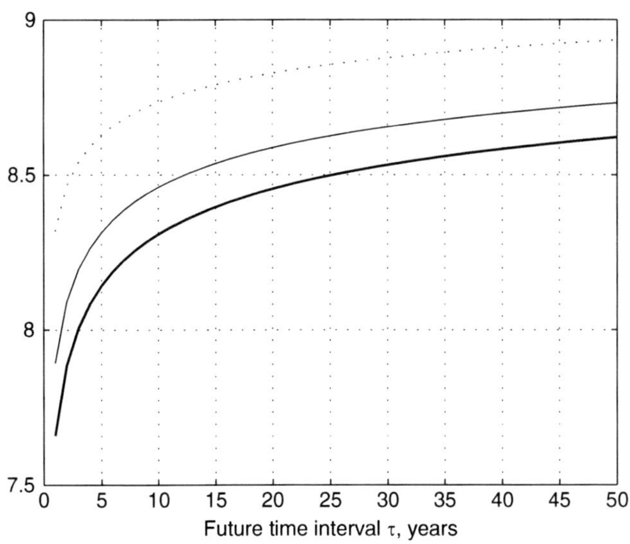

Fig. 6.16 The GEV-quantiles of the future τ-maxima, Japan. 99% (*top, dotted line*), 95% (*middle, light line*), 90% (*bottom, heavy line*)

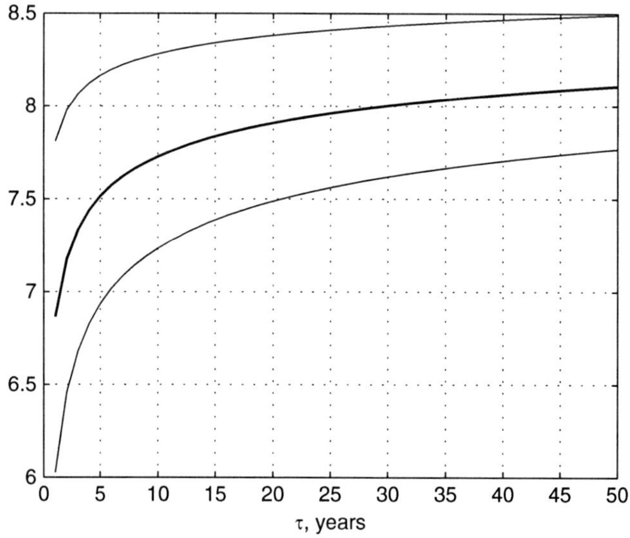

Fig. 6.17 The median (*middle*), 95% quantile (*top*), 5% quantile (*bottom*) of the future τ-maxima, Japan. The GEV fitting

6.5.2 Fennoscandia Catalog

In this section we apply our methods to the Magnitude Catalog of Fennoscandia [UP], which was provided by ICD-B2. The catalog covers the area $46.4 \le$ latitude $\le 85.8°$; $-24.9 \le$ longitude $\le 57.9°$ and the time period 01.01.1375 – 31.12.2005. Focal depth varies in the interval $0 \le$ depth ≤ 97.0 km; magnitude varies in the interval $-0.7 \le$ magnitude ≤ 6.6. There are 15,230 events in this time-space volume. The area includes Iceland and the mid-Atlantic ridge. Since we are interested in earthquakes occurring only in the Fennoscandia peninsula, we restrict our area of study by the polygon with coordinates

Latitude	Longitude
72.0	20.0
72.0	40.0
70.0	40.0
50.0	10.0
55.0	0
60.0	−5.0
65.0	0

The contour map of this area is shown in Fig. 6.18. The earthquake with maximum magnitude 5.8 occurred in this area on 01.09.1819 at the point $\lambda = 66.4$; $\phi = 14.4$; depth unknown (Norway). Since we are interested in the analysis of main shocks, we have applied the Knopoff-Kagan algorithm described above to

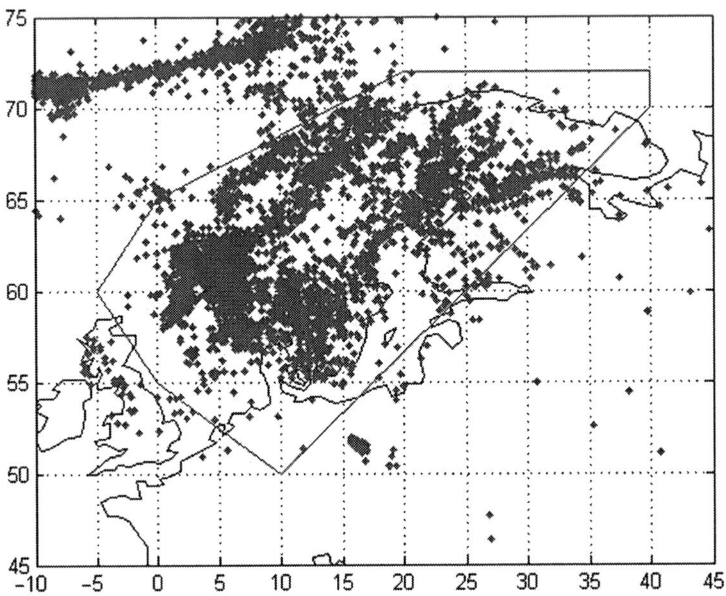

Fig. 6.18 The contour map of the Fennoscandia region with the polygon of study and earthquake epicenters, 1900–2005

remove aftershocks from the catalog. We got the following numbers of main shocks:

$h = -0.7$ 6868 main shocks; intensity $\lambda = 0.1774$ events/day $= 64.70$ events/year

$h = 2.7$ 1576 main shocks; intensity $\lambda = 0.0407$ events/day $= 14.9$ events/year

$h = 3.3$ 594 main shocks; intensity $\lambda = 0.0154$ events/day $= 5.61$ events/year

The intensity of the 1900–2005 main shocks for these three thresholds in a moving 5-year window is shown in Fig. 6.19 (each intensity value is plotted versus the window center). It should be noted that the percentage of aftershocks is very low for Fennoscandia, about 6%, whereas for the other seismic zones it reaches 50% and even more. We see that the positive trend of intensity for $h = -0.7$ and $h = 2.7$ is almost invisible for $h = 3.3$. Figure 6.20 shows a histogram of main shocks. We see that the histogram becomes more or less regular and decreasing for $m \geq 2.2$. Perhaps there is some tendency toward half-integer values, but not a very prominent one. Figure 6.21 shows the sample tail $1 - F(x)$ for the 1900–2005 main shocks. We see that for $m \geq 2.2$ a linear dependence is clearly seen, although after $m = 4.0$ a somewhat steeper slope appears, which tells us about a faster decay of the magnitude PDF in this range.

To sum up, we can take for the analysis a more or less stationary sequence of events in the magnitude range $m \geq 3.3$. Above this threshold there are $n = 594$ events.

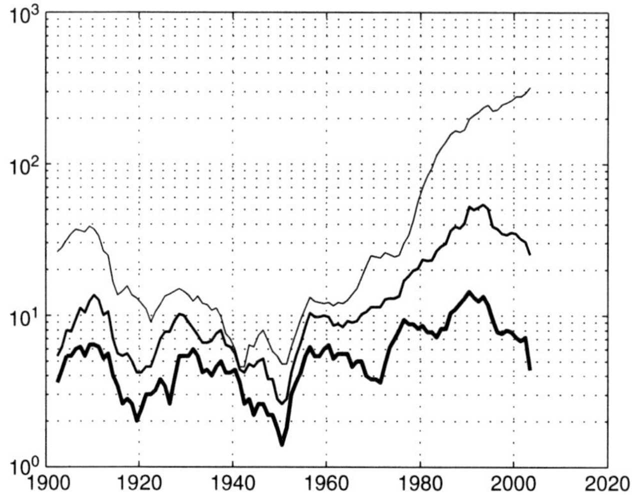

Fig. 6.19 The intensities (1/year) for the Fennoscandia catalog. Thresholds: $h = -0.7$ (*top*, $n =$ 6,868); $h = 2.7$ (*middle*, $n =$ 1,576); $h = 3.3$ (*bottom*, $n = 594$)

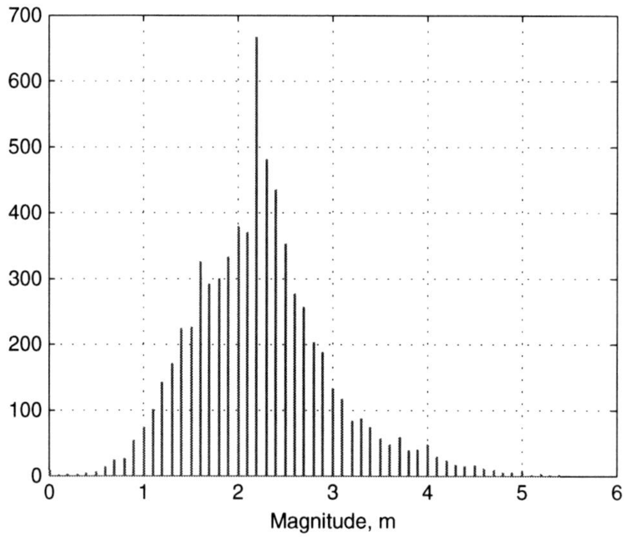

Fig. 6.20 The histogram of main shocks, Fennoscandia, 1900–2005, $n =$ 6,868

6.5.2.1 The GPD Fitting

We apply the GPD fitting to the Fennoscandia catalog. First of all, we have to specify the bins for the chi-square test. The histogram of main events is shown in Fig. 6.20. One can observe some irregularities in the range under analysis ($m \geq 3.3$), besides a non-monotonic behavior of the magnitude histogram. These irregularities

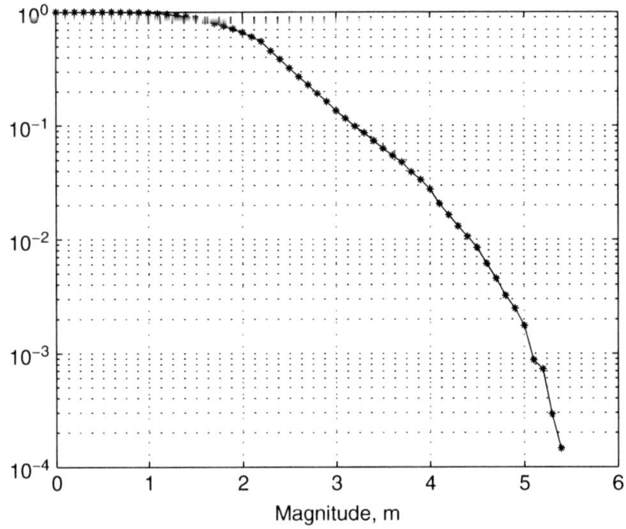

Fig. 6.21 The complementary sample distribution function $1 - F(x)$ of main shocks, Fennoscandia, 1900–2005, $m \geq -.7$; $n = 6{,}868$

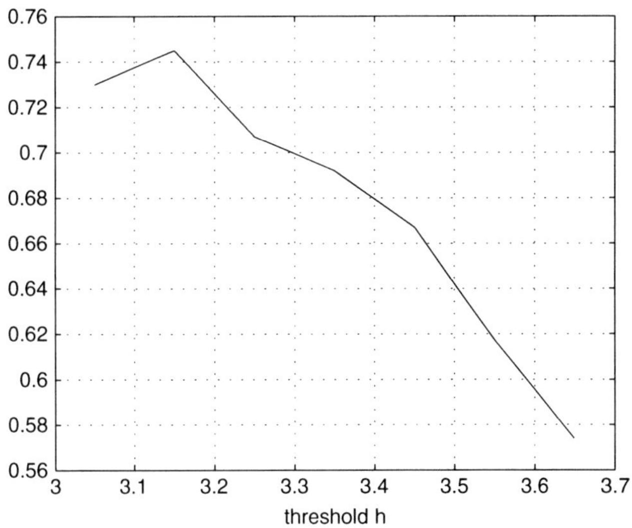

Fig. 6.22 The s-estimates for different thresholds by GPD-fitting. Fennoscandia, 100 bootstraps

forced us to combine the 0.1 bins into 0.2 bins. The use of 0.2 bins must be sufficient to remove or at least to diminish the irregularities.

Figure 6.22 shows the estimates of the parameters s for different thresholds h. As can seen in Fig. 6.22, the estimates begin decreasing in accordance with the theoretically expected tendency (6.10) at $h \geq 3.15$. The range under analysis

Table 6.3 Estimates provided by the chi-square fitting procedure using the GPD approach

h	3.25	3.35	3.45	3.55
r	7	7	7	7
Degrees of freedom	5	5	5	5
ξ	−0.297	−0.315	−0.299	−0.268
s	0.705	0.692	0.667	0.617
M_{max}	5.63	5.46	5.57	5.76
$Q_{0.90}(10)$	5.23	5.20	5.27	5.35
P_{exc}	0.11	0.13	0.19	0.12

The parameters are estimated by minimizing $S(\xi,s)$, as given by (6.57). M_{max} is the rightmost point of the magnitude distribution given by (6.6). $Q_{0.95}(10)$ is the 95% quantile of the maximum magnitude distribution (T-maximum magnitude) in 10-year intervals

($m \geq 3.3$) thus satisfies the requirement of a monotone decrease of s. As mentioned above, the number of discrete bins for the chi-square test should be as great as possible, in particular in the farther tail, provided the number of observations in each bin is no less than 8. By several numerical experiments we determined that the best number of 0.2 bins is 7, whereas the last, eighth non-standard bin which contains the tail as far as M_{max}, can have a width different from 0.2. For thresholds $h \geq 3.65$ the data were scarce, and the number of observations per bin is less than 8 in the tail. Thus, we have to restrict the thresholds from above by this value and to analyze the estimation results for the interval of thresholds $3.25 < h < 3.65$. We show in Table 6.3 the results of this estimation for $h = 3.25$; 3.35; 3.45; 3.55. For each threshold we derived the chi-square estimates that minimize the sum (6.57). We again used 100 bootstraps for each threshold value.

In Table 6.3, all the four thresholds give rather similar estimates. The significance levels of the chi-square sums are not much different, and none of them is critical. This suggests averaging the parameters ξ, $Q_{0.95}(10)$ for these thresholds, as suggested in Section 6.4. The resulting averaged values are

$$< \xi_{GPD} > \; = -0.295; \qquad < Q_{0.95}(10) > \; = 5.26. \tag{6.68}$$

Since the theoretical parameter s depends on h through Eq. 6.10, we converted the s-estimates for thresholds $h = 3.35$; 3.45; 3.55 into estimates of $s(h = 3.25)$:

$$\begin{aligned}
0.692 - &< \xi_{GPD} > \cdot (3.35 - 3.25) = 0.7215;\\
0.667 - &< \xi_{GPD} > \cdot (3.45 - 3.25) = 0.7260;\\
0.617 - &< \xi_{GPD} > \cdot (3.55 - 3.25) = 0.7055.
\end{aligned}$$

Now we get the mean value:

$$< s_{GPD}(h = 3.25) > \; = (0.705 + 0.7215 + 0.7260 + 0.7055)/4 = 0.7145; \tag{6.69}$$

Using Eq. 6.6 and estimates (6.68), (6.69), we get an estimate of M_{max}:

$$< M_{max} > \; = 3.25 - \; < s_{GPD}(3.25) > / < \xi_{GPD} > \; = 5.67. \tag{6.70}$$

In order to estimate the statistical scatter of the estimates, we simulated our whole procedure of estimation 1,000 times by generating GPD-samples with parameters (6.68), (6.69). As well as for the Japanese catalog, we characterized the scatter by sample standard deviations, since all estimates but those of M_{max} had light tails and small skewness values. As to the estimates of M_{max}, their distribution was skewed towards larger values. Combining the std estimates derived from simulations with the mean values (6.69), (6.70) based on the actual catalog, the final results of the GPD approach for the Fennoscandia catalog can be summarized by

$$\xi_{GPD} = -0.295 \pm 0.025; \quad s_{GPD}(3.25) = 0.715 \pm 0.029; \tag{6.71}$$

$$M_{max,\,GPD} = 5.67 \pm 0.165; \quad Q_{0.95}(10) = 5.26 \pm 0.073. \tag{6.72}$$

$$quantiles\ of\ M_{max} : q_{0.16} = 5.31; \quad q_{0.84} = 6.07. \tag{6.73}$$

One can observe that the std of M_{max} exceeds the std of the quantile $Q_{0.95}(10)$ by a factor greater than two, confirming once more our earlier conclusion as to a larger instability of M_{max} compared with $Q_{0.95}(10)$.

6.5.2.2 The GEV Fitting

We now apply the GEV fitting to the Fennoscandia catalog. The intensity of the Poisson process is $\lambda = 0.0154$ events per day. In accordance with the GEV-restriction (1), the T-interval should satisfy the inequality

$$T > 7/\lambda \cong 450.$$

On the other hand, the GEV-restriction (2) requires that

$$T < T_{cat}/60 \cong 644.$$

Thus, in this situation we have a rather narrow interval for acceptable values of T inside of (450; 650). So we took this interval for the estimation of GEV-parameters. A 100 shufflings with subsequent averaging were used. The results are given in Table 6.4.

In Table 6.4, all three T-intervals provide rather similar estimates. The significance levels of the corresponding chi-square sums are not much different and are about 0.50. This suggests averaging the parameters for these thresholds. We get averaged values:

$$\xi_{GEV} = -0.308; \quad M_{max,\,GEV} = 5.88; \quad Q_{0.95}(10) = 5.51; \tag{6.74}$$

Table 6.4 Estimates provided by the chi-square fitting procedure using the GEV approach

T	450	550	650
R	6	6	6
Degrees of freedom	4	4	4
ξ	−0.297	−0.315	−0.313
μ	4.425	4.523	4.609
σ	0.466	0.443	0.417
M_{max}	5.92	5.85	5.88
$Q_{0.90}(10)$	5.48	5.45	5.48
P_{exc}	0.51	0.52	0.54

The parameters are estimated by minimizing $S(\xi,\mu,\sigma)$ as given by (6.63). M_{max} is the rightmost point of the magnitude distribution given by (6.4). $Q_{0.95}(10)$ is the 95% quantile of the maximum magnitude distribution (T-maximum magnitude) in 10-year intervals

Since the theoretical parameters μ, σ depend on T through (6.11), (6.12), we converted μ and σ estimates for $T = 550$; 650 into estimates for $T = 450$ using (6.15), (6.16):

$$\mu(T = 550) : \ 4.523 - (0.443/0.308) \cdot \left((450/550)^{-0.308} - 1 \right) = 4.431;$$

$$\mu(T = 650) : \ 4.609 - (0.417/0.308) \cdot \left((450/650)^{-0.308} - 1 \right) = 4.441;$$

$$\sigma(T = 550) : 0.443 \cdot (450/550)^{-0.308} = 0.471;$$

$$\sigma(T = 650) : 0.417 \cdot (450/650)^{-0.308} = 0.467.$$

Now we get mean values:

$$< \mu_{GEV}(T = 450) > \ = (4.425 + 4.431 + 4.441)/3 = 4.432; \tag{6.75}$$

$$< \sigma_{GEV}(T = 450) > \ = (0.466 + 0.471 + 0.467)/3 = 0.468. \tag{6.76}$$

The probability of exceeding the observed chi-square sum was estimated as $P_{exc} = 0.50 \div 0.54$ from 1,000 generated catalogs that obey the GEV-distribution with parameters (6.74), (6.75), and (6.76), so that the goodness of fit by the GEV distribution was quite acceptable. In order to estimate the statistical scatter of the estimates, we simulated our whole procedure of estimation 1,000 times by generating GEV-samples with parameters (6.74) (6.75), and (6.76). Combining the std estimates derived from simulations with the mean values (6.74) (6.75), and (6.76), the final results of the GEV approach for the Fennoscandia catalog can be summarized by the estimates:

$$\xi_{GEV} = -0.308 \pm 0.042; \quad M_{max,\,GEV} = 5.89 \pm 0.31; \quad Q_{0.95}(10) = 5.51 \pm 0.15; \tag{6.77}$$

$$\mu_{GEV}(T=450) = 4.432 \pm 0.0402; \quad \sigma_{GEV}(T=450) = 4.432 \pm 0.0414; \quad (6.78)$$

$$quantiles\ of\ M_{max} : q_{0.16} = 5.75; \quad q_{0.84} = 6.37. \quad (6.79)$$

One notes that the std of M_{max} exceeds the std of the quantile $Q_{0.95}(10)$ by a factor greater than two, confirming once more our earlier conclusion as to a larger instability of M_{max} compared with $Q_{0.95}(10)$. Finally, we show in Fig. 6.23 the quantiles $Q_q(10)$, $q = 0.90$; 0.95; 0.99 as functions of τ for $\tau = 1 \div 50$ years obtained by the GPD approach. Figure 6.24 plots the median of $Q_q(10)$ together with two accompanying quantiles 5% and 95% which enclose the 90% confidence interval.

To compare the performances of the GEV and GPD approaches we collected in Table 6.5 estimates of the parameters independent of the choice of T-intervals and thresholds h along with their std.

This comparison between the performances of the GEV and the GPD approaches suggests that they are not so very different. If only the standard deviations are considered, the GPD approach is slightly better for Fennoscandia, whereas the GEV approach is a little better for Japan. The GEV approach provides somewhat higher estimates of M_{max} and $Q_{0.95}(10)$ for Japan and Fennoscandia. For the global Harvard catalog both approaches give similar estimates and std. Thus, we can conclude that the larger space-time volume of the global catalog ensured more stable estimation of the tail distribution parameters as compared with regional catalogs.

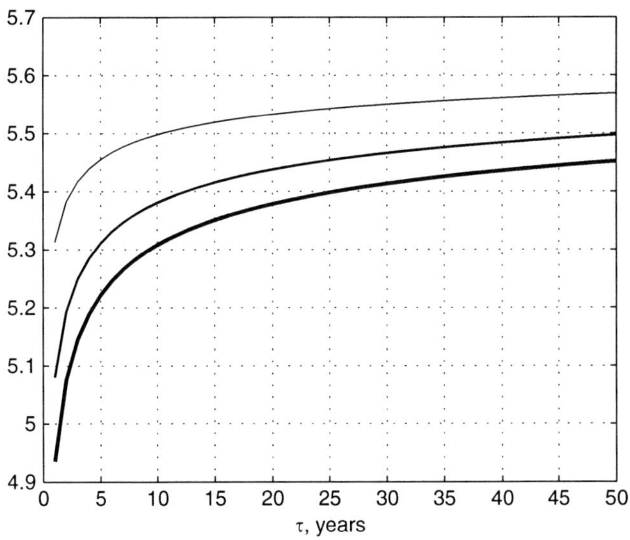

Fig. 6.23 The GPD-quantiles of the future τ-maxima, Fennoscandia. 99% (*top*), 95% (*middle*), 90% (*bottom*)

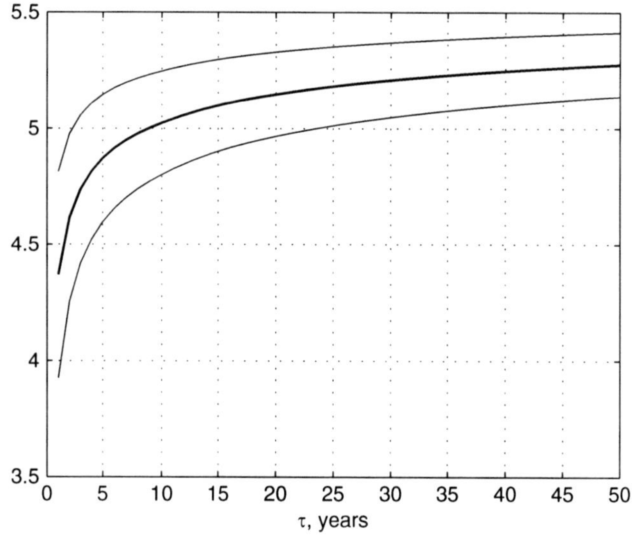

Fig. 6.24 84% (*top*), 50% (*middle*), 16% (*bottom*) quantiles of the maximum magnitude distribution for a future time interval τ (year). Fennoscandia, the GPD-fitting

Table 6.5 Comparison of performance between the GEV and GPD approaches based on three catalogs: the global Harvard catalog, the catalog of Japan, and the catalog of Fennoscandia. Statistical estimates derived by the GEV and GPD approaches are shown along with their std

Catalog	Harvard	Japan	Fennoscandia
GPD approach			
Shape parameter ξ	-0.197 ± 0.054	-0.214 ± 0.103	-0.295 ± 0.025
M_{max}	9.54 ± 0.55	9.31 ± 1.14	5.67 ± 0.165
$Q_{0.95}(10)$	8.89 ± 0.23	8.29 ± 0.49	5.26 ± 0.073
GEV approach			
Shape parameter ξ	-0.190 ± 0.060	-0.190 ± 0.072	-0.308 ± 0.042
M_{max}	9.54 ± 0.65	9.57 ± 0.86	5.89 ± 0.210
$Q_{0.95}(10)$	8.88 ± 0.21	8.34 ± 0.32	5.51 ± 0.102

6.6 Main Results

In comparing the global Harvard catalog with regional catalogs of Japan and Fennoscandia, we note one essential difference. The global catalog includes magnitudes that are called "outliers". At the same time, the tail of the magnitude distribution in the catalogs of Japan and especially Fennoscandia dies out more rapidly. Thus, one can say that regional catalogs are "more compact". Accordingly, one can expect that magnitudes with significantly higher values than the observed ones are hardly possible in Fennoscandia, even over a long time. The question concerning the causes of this difference is still under discussion. Perhaps the phenomenon may be due to a high heterogeneity of the area covered by the global catalog and much longer time intervals of the regional magnitude catalogs. The parameter estimates for the global Harvard catalog are less stable in general than those for the regional catalogs, which also can be explained by the same reasons.

The physical constraints upon the maximum size of earthquakes (a consequence of the finiteness of the Earth) corresponds to the condition $\xi < 0$, whereas the ordinary Gutenberg–Richter law (for magnitudes) corresponds to the value $\xi = 0$ (unlimited exponential distribution).

Our estimates correspond to the first case, where the values ξ close to –0.2 have been obtained for the global catalog and $-0.2 \div -0.3$ for the regional catalogs. These values correspond to a certain upper boundary M_{max} in the magnitude distribution. However, statistical estimates of the parameter M_{max} are quite unstable, in particular, when the absolute value of the shape parameter ξ is low. That is why we propose, instead of the M_{max} parameter, to use the stable quantiles $Q_q(\tau)$.The quantiles $Q_q(\tau)$ can be very useful tools for pricing risks in the insurance business and for optimizing the allocation of resources and preparedness by governments.

It must be emphasized that the fitting by the GEV or by GPD does not a priori assume that the distribution is either limited or unlimited. The character of the distribution (restricted or otherwise) is determined by the sign of ξ. Our estimates of this parameter clearly show that the parameter ξ is negative with a high probability.

The results we obtained provide a better understanding of how the tail of the distribution of earthquake size (moment, energy) behaves. Different authors have tried to describe the distribution of seismic moments by unlimited functions: the gamma function, the exponential function with a fractional index (the stretched exponential distribution), and the Pareto distribution with an index higher than unity. The results obtained ($\xi < 0$) indicate that the modeling of the distribution of earthquake size by a law limited from the right is more adequate.

We have adapted the new method of statistical estimation of tail parameters to deal with earthquake catalogs, including those involving discrete magnitudes. The method is based on the duality of the two main limit theorems in Extreme Value Theory (EVT). One theorem leads to the GPD (peaks over threshold approach), the other to the GEV (T-maximum method). Both limiting distributions must possess the same shape parameter ξ. Both approaches provide almost the same statistical estimate for the shape parameter, which is negative. A negative shape parameter corresponds to a distribution of magnitudes which is bounded from above (by a parameter called M_{max}). This maximum magnitude stems from the finiteness of the geological structures that generate earthquakes. The density distribution extends to its rightmost value M_{max} with a very small probability mass in its neighborhood characterized by a tangency of a high degree ("duck beak" shape). In fact, the limit behavior of the density distribution of earthquake magnitudes is described by the function $(M_{max} - m)^{-1-1/\xi} \cong (M_{max} - m)^4$, i.e., by a polynomial of degree approximately equal to 4. This explains the instabilities in the statistical estimates of M_{max}: a small change in the catalog of earthquake magnitude can give rise to a significant fluctuation of the resulting estimate of M_{max}. In contrast, the estimation of the "integral" parameter $Q_q(\tau)$ is generally more stable and robust, as we have demonstrated above.

The main problem in the statistical study of the tail for a distribution of earthquake magnitudes (and this applies to distributions of other rarely observable extremes) is the estimation of quantiles which are beyond the data range, i.e., quantiles of level $q > 1 - 1/n$, where n is the sample size. We would like to stress once more that reliable estimation of quantiles of levels $q > 1 - 1/n$ can be made only with some additional assumptions on the behavior of the tail. Such assumptions can sometimes be made on the basis of physical processes underlying the phenomena under study. We used for this purpose general mathematical limit theorems, namely, theorems of the Extreme Value Theory (EVT). In our case, the assumptions for validity of the EVT boil down to assuming a regular (power-like) behavior of the tail $1 - F(m)$ in the distribution of earthquake magnitudes in the vicinity of its rightmost point M_{max}. Some justification for such an assumption can be the fact that, without it, there is no meaningful limit theorem in the EVT. Of course, there is no a priori guarantee that these assumptions will hold and the relevant methods would work well in some concrete situation. In fact, because the EVT suggests a statistical methodology for the extrapolation of quantiles *beyond* the data range, the question whether such interpolation is justified or not in a given problem should be investigated carefully in each concrete situation. But the EVT provides the best statistical approach possible in such a situation.

Appendix A: Application of the Kolmogorov Test to the Densities That Depending on a Parameter

The quality of the fit of a theoretical distribution function $F(x/\alpha)$ that depends on a parameter α (α may be multidimensional) to an empirical distribution can be tested by some goodness-of-fit metric. We have chosen the Kolmogorov distance KD:

$$KD = n^{1/2} max|F(x|\hat{\alpha}) - F_n(x)|, \tag{6.80}$$

where $\hat{\alpha}$ is a statistical estimate of α, e.g., the maximum likelihood estimate, or the moment-estimate. $F_n(x)$ is the sample stepwise distribution function which is defined as follows. If we denote the ordered sample as $x_1 \leq \cdots \leq x_n$, then

$$F_n(x) = \begin{array}{ll} 0; & x \leqslant x_1; \\ m/n; & x_m < x \leqslant x_{m+1}; \\ 1; & x > x_n. \end{array} \tag{6.81}$$

Since we use a theoretical function with parameters fitted to the data, we cannot use the standard Kolmogorov distribution to find the significance level of the KD. Instead, in order to determine the significance level of a given KD-distance based on a given sample, one has to use a numerically calculated distribution of KD-distances, measured in the simulation procedure with fitted parameters. This method was suggested in [St] for the Gaussian and the exponential distributions.

We use the Kolmogorov distance to test the admissibility of the GEV distribution for T-maxima, and the GPD distribution for "peaks over threshold" magnitudes, i.e., magnitudes exceeding a threshold h.

Appendix B: Estimation of the Parameters (μ,σ, ξ) of the GEV Distribution Function: The Method of Moments (*MM*)

Consider a sample $(x_1,\ldots,x_n)$ drawn from the distribution function $GEV(x\,/\mu,\sigma,\,\xi)$:

$$GEV(x|\mu,\sigma,\xi) = exp(-[1+(\xi/\sigma)\cdot(x-\mu)]^{-1/\xi});\quad 1+(\xi/\sigma)\cdot(x-\mu)>0;\quad \xi<0. \tag{6.82}$$

We shall solve the problem of estimating the parameters (μ,σ,ξ) by the *Method of Moments* (*MM*). Since the support of the distribution $GEV(x)$ depends on the unknown parameters as defined in (6.82), the regularity of the *Maximum Likelihood Method* (*MLE*) and its asymptotic efficiency properties are not obvious. It was shown that the *MLE* are asymptotically regular at least for $\xi > -0.5$. But even with this restriction the *MLE* turned out to be less efficient than the *MM* for small and moderate sample sizes ($n \cong 10 \div 100$) (see [Ch, CD, PSSR]). Consequently, we prefer to use the *MM*.

Let us recall the definitions of the *MM*. The first three centered sample moments are denoted as follows:

$$M_1 = \frac{1}{n}\sum_{k=1}^{n} x_k;\quad M_2 = \frac{1}{n}\sum_{k=1}^{n}(x_k - M_1)^2;\quad M_3 = \frac{1}{n}\sum_{k=1}^{n}(x_k - M_1)^3. \tag{6.83}$$

The corresponding theoretical moments for the *GEV*-distribution (6.82) are

$$m_1 = \mu - \sigma/\xi + (\sigma/\xi)\cdot Gamma(1-\xi); \tag{6.84}$$

$$m_2 = \sigma^2/\xi^2[Gamma(1-2\xi) - (Gamma(1-\xi))^2];$$

$$m_3 = \sigma^3\cdot[-2(Gamma(-\xi))^3 - (6/\xi)\cdot Gamma(-\xi)Gamma(-2\xi) - (3/\xi^3)\cdot Gamma(-3\xi)];$$

The *MM* identifies the theoretical moments with their sample analogues and solve the resulting equations for the unknown parameters:

$$\mu - \sigma/\xi + (\sigma/\xi)\cdot Gamma(1-\xi) = M_1; \tag{6.85}$$

$$\sigma^2/\xi^2[Gamma(1-2\xi)-(Gamma(1-\xi))^2]=M_2; \tag{6.86}$$

$$\begin{aligned}&\sigma^3\cdot[-2(Gamma(-\xi))^3-(6/\xi)\cdot Gamma(-\xi)Gamma(-2\xi)\\&\quad-(3/\xi^3)\cdot Gamma(-3\xi)]=M_3.\end{aligned} \tag{6.87}$$

We can solve *explicitly* Eqs. 6.85–6.86 for μ,σ:

$$\sigma=[(M_2\xi^2)/(Gamma(1-2\xi)-(Gamma(1-\xi))^2)]^{1/2}; \tag{6.88}$$

$$\mu=\sigma/\xi-(\sigma/\xi)\cdot Gamma(1-\xi)+M_1. \tag{6.89}$$

Inserting (6.88) into Eq. 6.89, we have only one 1D equation for the numerical search of the solution for ξ. This can be done using any standard numerical algorithms.

Appendix C: Estimation of Parameters (s,ξ) of the GPD by Maximum Likelihood (ML) Method

We denote a sample consisting of peaks exceeding a threshold h by $y_1,\ldots,y_N$, the number of peaks being dependent on h: $N=N(h)$. We assume that h is high enough and N is large enough so that, in accordance with Theorem 2, the sample distribution of $y_1,\ldots,y_N$ is satisfactorily fitted by a *GPD* with some parameters (s,ξ) to be estimated. The log likelihood function is

$$L(s,\xi)=-N\cdot log(s)-(1+1/\xi)\sum_{k=1}^{N}log(1+(\xi/s)\cdot(y_k-h)) \tag{6.90}$$

Maximization of the likelihood (6.90) which is a function of two parameters can be reduced to numerical maximization of a function that depends on a single parameter, which is much easier.

We define a new parameter as follows:

$$\theta=\xi/s. \tag{6.91}$$

Then we can rewrite the likelihood for parameters (θ,ξ):

$$L(\theta,\xi)=-N\cdot log(\xi/\theta)-(1+1/\xi)\sum_{k=1}^{N}log(1+\theta\cdot(y_k-h)). \tag{6.92}$$

The maximum of likelihood (6.92) over parameter ξ is found explicitly. It is provided by

$$\xi = \xi(\theta) = (1/N) \cdot \sum_{k=1}^{N} log(1 + \theta \cdot (y_k - h)). \tag{6.93}$$

Inserting (6.93) into (6.92), we get the likelihood L(θ) depending on the single parameter θ:

$$L(\theta) = - N \cdot log\left((1/N) \cdot \sum_{k=1}^{N} log(1 + \theta \cdot (y_k - h)) \right) + N \cdot log(\theta) - \sum_{k=1}^{N} log(1 + \theta \cdot (y_k - h)) - N. \tag{6.94}$$

Now (6.94) can easily be maximized over the single variable θ by standard numerical algorithms. Denote the resulting value by θ*. Returning to the initial parameters (s,ξ) through (6.91), we get their *ML* estimates ξ*, s*:

$$\xi^* = \xi(\theta^*); \quad s^* = \xi(\theta^*)/\theta^*. \tag{6.95}$$

This method works fine if $\xi > -1/2$. In this case the asymptotic distribution of the pair

$$(\xi^* - \xi) \cdot N^{1/2}; (s^*/s - 1) \cdot N^{1/2}$$

is normal with zero mean and covariance matrix C:

$$C = \begin{matrix} (1 + \xi)^2 & -(1 + \xi) \\ -(1 + \xi) & 2(1 + \xi) \end{matrix}. \tag{6.96}$$

The usual MLE properties of consistency and asymptotic efficiency hold (see [EKM]).

Chapter 7
Relationship Between Earthquake Losses and Social and Economic Situation

tu ne cede malis
Sed contra audentior ito!

Vergilius

do not recede before disaster
But bravely go against it!

Virgil

An analysis of the average amount of losses from natural disasters both in the death toll and in economic loss leads to the conclusion about a fast nonlinear growth of losses with time. This tendency used to be ascribed to a population growth, the development of potentially hazardous industries, and a general deterioration of the environment. Extrapolation of this tendency suggests that all the economic gain will be consumed in the greater losses from natural disasters by the end of the twenty-first century (or even by the mid-twenty-first century). However, this pessimistic prediction is based on the use of means and variances, which is incorrect for the case of heavy-tailed distributions typical of death toll and economic loss from natural disasters. The application of statistically correct approaches affects the prediction essentially.

7.1 Variation in the Number of Casualties and Economic Loss from Natural Disasters

Mass media abound in information about natural disasters: hurricanes, earthquakes, floods, droughts occur regularly over the globe. Consequences of such disasters could be horrific. The death toll of the great 2004 Sumatra earthquake amounted to 227,898 people, in the 1976 Tangshan earthquake in China, according to different sources, there were as many as 240,000–650,000 casualties. The floods of 1931 and 1970 in China and Bangladesh took 1,300 and 500,000 human lives, respectively.

Many authors have studied tendencies in the variation in the death tolls and economic losses from natural disasters. Based on the data reported at the World

V. Pisarenko and M. Rodkin, *Heavy-Tailed Distributions in Disaster Analysis*, Advances in Natural and Technological Hazards Research 30,
DOI 10.1007/978-90-481-9171-0_7, © Springer Science+Business Media B.V. 2010

Conference on Natural Disasters held in Yokohama (1994), the number of disasters over the period 1962–1992 increased by a factor of 4.1, the losses increased on the average by 6% per year, and the number of casualties increased on average by 4.3% [VMO]. Based on these figures, the unfavorable forecast was got that by the end of the twenty-first century (or even by the mid-twenty-first century) all economic gain will be taken up by the growing losses from natural disasters [O2, O3]. Figure 7.1 provides examples of this tendency inferred from global data on the number of major disasters and the associated losses (Interternational Disaster Data-base, http://www.em-dat.net, accessed in 2006). This tendency of disaster growth both in number and in loss values from disasters used to be ascribed to increasing population, the development of potentially hazardous industries, a general deterioration of the environment, and to the involvement in the operating economy of less environmentally friendly territories.

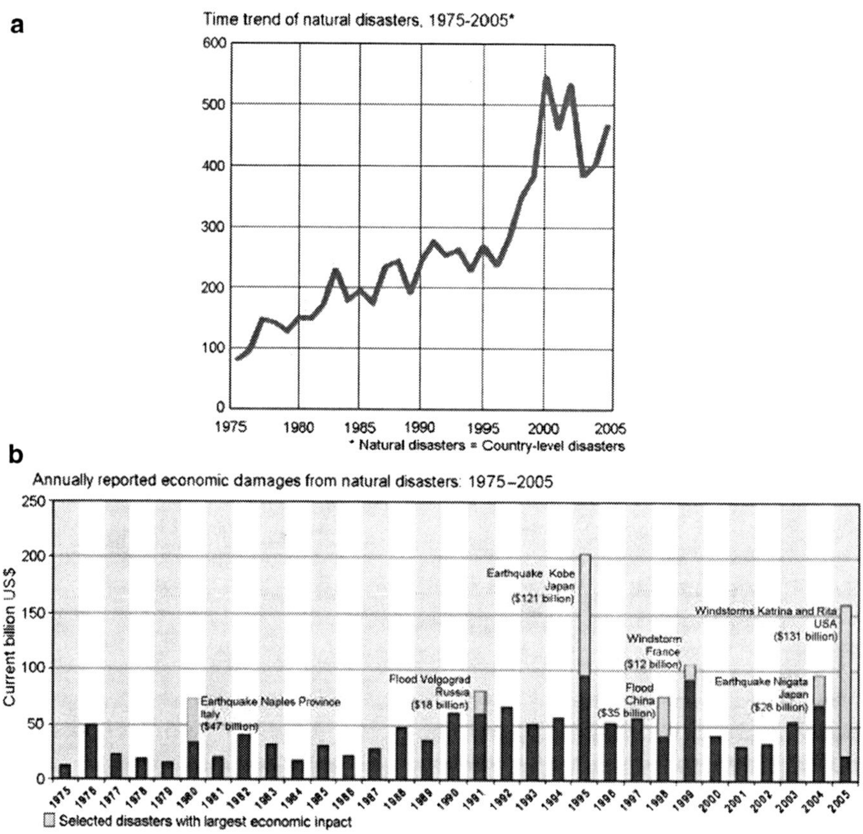

Fig. 7.1 Change in the number of natural disasters recorded in the world (a) and losses from disasters (b) (including individual most damaging events) during the period 1975–2005 based on EM–DAT (The figure was borrowed from http://www.em-dat.net, accessed in 2006)

However, the statistical approach used in this analysis is based on means and variances and hence, as shown in the previous chapters, the predictions obtained in this way are not quite correct. The problem here lies in the existence of rare occasional, anomalously great losses comparable with the total loss. Actually, the death toll from the two abovementioned – Tangshan and Great Sumatra – earthquakes amounts to about one-third of the casualties from all earthquakes for the entire twentieth century. Such a comparison makes one think that the conventional approach based on the calculation of average values cannot give robust estimates; hence other statistical approaches presented in the previous chapters are required to study the loss pattern from disasters. A correct analysis of the loss pattern is given below and, as a result of this analysis, the prediction changes drastically. Instead of predicting of a disastrous growth of losses with time, the model of quasi-stationary level of relative losses that even tend to decrease with socio-economic development seems to be better grounded.

The study presented below is based mainly on the database of earthquake losses described in [RP1, RP3, RP5] and in Chapter 4 in detail. The choice was due to the fact that the data allowing to carry out the examination of change in the death toll from natural disasters over the last century are available only for earthquakes; moreover, starting from the 1950 to 1960, the data on economic losses are available with increasing completeness also. As to the other natural disasters, the available information is substantially less satisfactory. However, the data available convincingly imply a qualitative uniformity of loss patterns from different kinds of disasters, the heavy-tailed distributions are found to be typical of different types of disasters, and consequently, the conclusions we draw are valid not only for the case of earthquake disasters but are applicable also to other kinds of natural disasters.

The statistical tendency of a nonlinear pattern of growth of cumulative losses typical of the case of heavy-tailed distributions was examined in detail in the previous Chapters 4 and 5. Let us sum up and memorize the main results. The analysis of the power law distributions implies that a nonlinear growth of cumulative losses can occur in a stationary model, i.e., the empirically observed nonlinear growth of loss values does not necessarily imply a non-stationarity in the disaster pattern. In the case of earthquakes, it turns out that the observed tendency of loss growth with time might be attributed to the heavy-tailed character of the loss distribution from earthquakes uniformly distributed through time. Factually, as it was shown above, the death toll from earthquakes can be described by the heavy tail distribution law with power parameter $\beta = 0.7$. Thus, the expected mean nonlinear growth of the death toll can be evaluated by the exponent $1/\beta = 1/0.7 = 1.4$, whereas the empirical exponent of average growth is equal to 1.6. Thus, the observed growth of the number of casualties caused by earthquakes appears to be in accordance with the stationary model if the heavy-tailed character of loss distribution is taken into account. The actual temporal change in the seismic disaster pattern is of secondary importance.

Naturally, in the case of other kinds of natural disasters, for example, hurricanes or floods, whose recurrence rate may depend essentially on the climate change, a real non-stationarity may be of greater importance. However, at least partly, the observed effect of nonlinear growth of loss will be caused by the power-law distribution of losses, and this part of the loss growth will occur even in the case of uniform disaster distributions in time. The results supporting this idea for the case of losses from hurricanes will be presented below. For the number of people who became homeless due to floods a key role of this factor is shown in [P4].

There is however an evident limitation, the death toll from any disaster cannot exceed the population of the Earth, and any loss of property cannot exceed the total cost of the technosphere. This means that the power law of loss distribution that is valid for moderate and large hazards cannot be valid also for the case of greatest loss values. The real pattern of loss distribution will not satisfy the power-law distribution with $\beta \leq 1$ for events $x > A$, where A is a certain characteristic value. An event with a size $x \cong A$ is called the characteristic maximum possible disaster (see Chapter 5 for details). For a time interval greater than that of recurrence of such disaster, the loss growth pattern will change in a qualitative way. The total losses will increase with time in a linear manner, and the use of mean values becomes justified. That means that it is important to evaluate the size of characteristic maximum event A and the recurrence time T_c of such events.

An approach used to estimate the maximum possible event A value and the recurrence period of such events T_c was presented and verified in Chapter 5 and in the papers [RP4, RP]. For more correct determination of these parameters the available data set should be subdivided into more uniform groups. Afterward we will discuss a relation between the statistics of natural disasters and socio-economic parameters. Having this in mind we will compare losses due to seismic disasters in different regions: North America, Europe, Japan, South America, Asia, and Indochina (Table 7.1). To evaluate the characteristic maximum possible disaster A value and its recurrence time T_c we will combine these regions into two groups according to higher (North America, Europe, Japan) and lower (South America, Asia,

Table 7.1 Comparison between characteristics of death toll and losses from earthquakes and socio-economic conditions in different countries

Region (1)	Number of events with known loss values (2)	Mean loss, $\$10^6$ (3)	Mean loss/casualties ratio, $\$10^6$ per capita (4)	Annual value of products, $\$10^3$ per capita (for 1970) (5)	(4)/(5) ratio $\times 10^3$ (6)	(3)/(5) ratio $\times 10^3$ per capita per annum (7)
North America	36	800	32	4.5	7.1	180
South Europe	167	340	8	1.5	5.3	230
Japan	74	430	5.5	1.6	3.4	270
Latin America	236	130	1.3	0.5	2.6	260
Asia	415	50	1	0.2	5	250
Indochina	133	30	1.2	0.2	6	150
max/min ratio	–	45	32	30	3	2.2

Indochina) level of economic development. This subdivision nearly coincides with subdivision into the groups of developed and developing countries.

To find changes in the disaster pattern due to rapid socio-economic development in the twentieth century, the data for the periods 1900–1959 and 1960–1999 were examined separately; the numbers of events within these two time intervals are almost the same. Assuming the death toll distribution is stable during these two time intervals, we estimate typical cumulative death toll $<V(t)>$ over time periods from 1 to 60 and from 1 to 40 years correspondingly for the first and second half of the twentieth century. Angle brackets denote the procedure of bootstrap averaging described in detail in Chapter 4 (we choose randomly yearly numbers of victims to model accumulation of death toll from 1 to t years , and then we take the medians of the obtained random samples, 100 random samples were shown to be enough for a such procedure).

Figure 7.2 shows simulation results strongly suggesting a change in the growth of cumulative number of casualties from earthquakes with time. The change from nonlinear growth law corresponding to the power-law distribution with $\beta < 1$ and with an infinite mean value to the linear growth law corresponding to some (unknown) distribution law with a finite mean value can be seen clearly.

The obtained typical cumulative loss values $<V(t)>$ were used to estimate A and T_c parameters characterizing the seismic loss patterns in developed and developing countries in the two time intervals (1900–1959 and 1960–1999). The procedure described in Chapter 5 was used. Growth of $lg<V(t)>$ values with time, and the change in growth pattern from the non-linear to the linear mode was approximated as

$$\lg(<V(t)>) = a_2 \lg^2(t) + a_1 \lg(t) + a_o, \qquad (7.1)$$

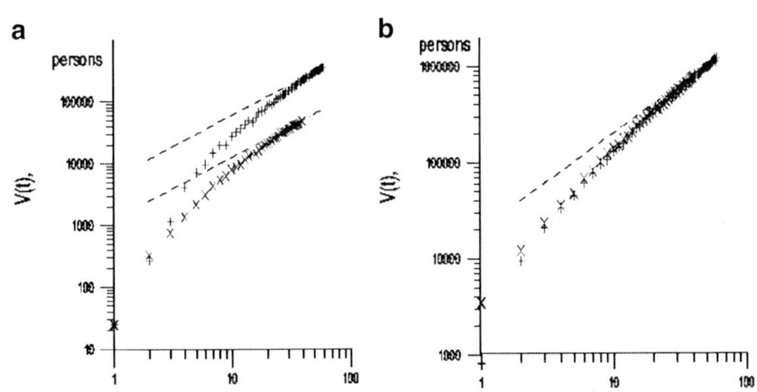

Fig. 7.2 Models of the growth of the typical death tolls $V(t)$ for developed (**a**) and developing (**b**) countries. Estimates are given for the periods of 1900–1959 (+) and 1960–1999 (×). *Dotted lines* are *straight lines* corresponding to a linear growth pattern of the death toll with time. The change in the growth pattern of the total death toll shows a transition from nonlinear to linear growth. There is also a tendency of decreasing death toll in the developed countries

where the parameters a_i ($i = 0, 1, 2$) in (7.1) can be evaluated by the least squares method. The time interval T_c needed for the change to the linear mode of $<V(t)>$ growth can be evaluated from the condition

$$d\{lg(<V(t)>)\}/d(lg(t))|_{t=Tc} = 1 \tag{7.2}$$

Thus, one has from (7.1) and (7.2)

$$2a_2 \times \lg(Tc) + a_1 = 1 \tag{7.3}$$

Relations (7.1, 7.2, and 7.3) permit us to evaluate the duration T_c of interval needed for the change from the non-linear to the linear mode of *V(t)* growth. Using $V(T_c)$ value we can determine the size of the maximum single event A occurring during the time interval T_c. For such evaluation we use the relation between the sum $S(t) = S_1 + S_2 + \ldots S_n$ and the single maximum $max\{S_i\}$ value valid for power-law heavy-tailed distributions with
$\beta < 1$ [P4]:

$$S(t) \cong max(S_i)/(1 - \beta) \tag{7.4}$$

In this case $max(S_i)$ is the maximum yearly number of casualties from seismic hazards occurring with recurrence time T_c and $S(t)$ corresponds to typical cumulative death toll during T_c years.

If necessary the relation (7.4) can be used twice: firstly for the determination of the maximum yearly number of victims during T_c years, and secondly for the evaluation of the number of casualties from the single maximum hazard occurring in T_c time interval.

Figure 7.2 and Table 7.2 show *A* and T_c parameter values obtained for the cases of seismic hazards occurring separately in the first and second parts of the twentieth century in the developed and developing countries. As can be seen, the characteristic cumulative death toll and the size of characteristic maximum

Table 7.2 The change in characteristic maximum death tolls from earthquakes and floods in different regions with time

Region	T_c, years	A, death toll, persons	Max actual disaster, death toll, persons
Earthquakes			
Developed countries, 1900–1959	33	95,000	110,000
Developed countries, 1960–1999	30	24,000	17,000
Developing countries, 1900–1959	40	270,000	200,000
Developing countries, 1960–1999	65	260,000	240,000 (650,000)[a]
Floods			
North America and European Union, 1950–1979[b]	15	15,00	650
North America and European Union, 1980–2005[b]	10	500	200

[a]Expert (probably overestimated) estimate of the death toll from Tangshan earthquake, China, 27.07.1976

[b]Data from the database EM-DAT: the International Disaster Database (http://www.em-dat.net)

seismic disaster A decreased substantially in the second half of the twentieth century in the developed countries.

Note that the typical cumulative death toll $<V(t)>$ values obtained for the second half of twentieth century for the developed and developing countries can be treated also as a prognosis of the death toll from earthquakes expected to occur in the developed and developing countries in the first half of the twenty-first century; this prognosis can be expected to be valid until the change in population density and in the technosphere become essential.

A similar procedure was used to examine the less representative data on flood losses obtained from the database EM-DAT (http://www.em-dat.net, accessed in 2008). The results which show the decrease in the death toll from the floods in the developed countries are given in Table 7.2. Both A and T_c decrease essentially in the second half of the twentieth century.

Now consider the relationship between the losses due to seismic disasters and socio-economic parameters. The population of the world is known to have increased four-fold during the twentieth century. However, the corresponding growth of number of disasters with the number of casualties greater than 100 persons has not been reported. It gives a basis to assume an existing of a factor compensating the potential growth of vulnerability of society because of increase of population. It would appear natural to attribute such compensation to the development of the technosphere. If this is the case, then having in mind the difference in technical development in the developed and developing countries, one should assume overcompensation and undercompensation in the cases of developed and developing countries, respectively.

Figure 7.3 shows the number of disasters with casualties more than 100 persons that took place in developed and developing countries. It presents also linear extrapolations of a number of such disasters inferred from the data over the period 1900–1939. In both cases of the developed and developing countries the actual number of major seismic disasters in the second half of the twentieth century differs from the expected extrapolated value. The actual number of major seismic disasters turned out to be less than would be expected in developed countries and slightly higher than that in developing countries. This conclusion strongly testifies for technogenic compensation of increase in the vulnerability of society in result of the population growth.

Similar results were obtained when we used data on smaller seismic disasters with the number of casualties from 11 to 100 persons. But this change could result also from the improvement in the recording of moderate seismic disasters in developing countries during the twentieth century, and for this reason these cases are less convincing.

7.2 Dependence of Losses on Per Capita National Product Values

Let us now compare data from different regions (Table 7.1).The relation between the loss characteristics and the socio-economic parameters can be revealed from the comparison of data on losses and those of per capita national product GDP values

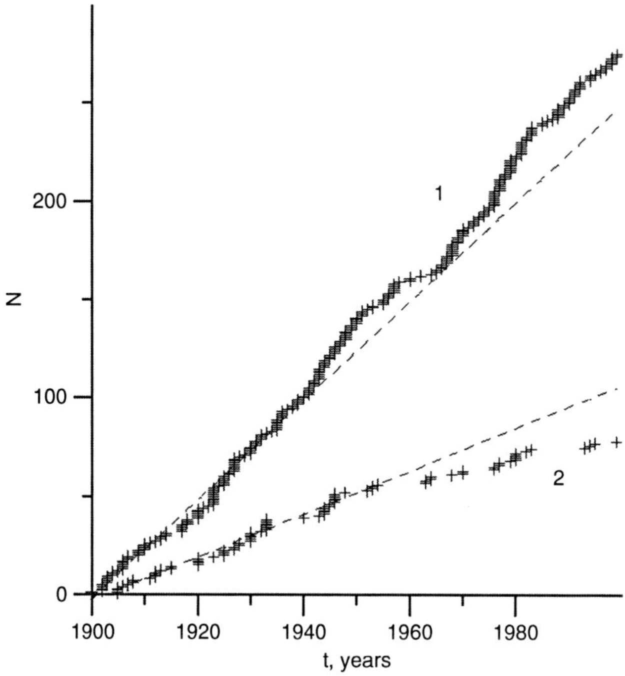

Fig. 7.3 Deviation of the number of disasters *N* with casualties above 100 persons from linear prediction (*dotted line*). (**1**) Developing countries, (**2**) developed countries. The disasters become less intense in the developed countries and slightly more intense in the developing countries

(obtained from the UNESCO database http://: unescostat.unesco.org/database, accessed in 2002). There is a correlation between the average ratio of economic loss per one life lost ("loss/casualties" ratio) and the national GDP. The "loss/casualties"/GDP ratio has a relatively constant value of 5.2 ± 2; minimum and maximum ratios differ only by a factor of 3, whereas the scatter of "loss/casualties" and GDP values being taken separately change more than 30 times. This means that material damage to the death toll ratio changes in the course of economic development, remaining roughly proportional to the GDP value.

The next topic to discuss is the patterns in economic loss from earthquakes. Economic earthquake losses are known to increase rapidly with time. However, to understand the change in the impact of disasters upon society one should envisage not this particular effect (fairly expected), but changes of relative losses norm to current level of national wealth. The GDP value is taken as the unit of loss values. The losses from major earthquakes in different regions measured in this unit turned out to be almost constant, namely (200 ± 50) thousands (column 7, Table 7.1). Such a result might be interpreted in terms of a quasi-stable level of relative economic loss from earthquakes for different countries despite essentially different seismic regime and different level of economic development.

The losses from seismic disasters increase greatly (and often are more accurately estimated) if earthquake hazards affect a megacity. To compare the loss patterns with the economic development we used data from [KN] on destructive earthquakes affecting large cities and occurring in 1971–1995. A required set of data (magnitude, population, number of casualties, economic loss, GDP value) was collected only for a moderate number of events. Figure 7.4 shows the dependence between economic loss per capita and GDP values. There is a close correlation between the parameter J – "loss/casualties" – and the annual per capita national product Q. The correlation coefficient $r = 0.94$. This result supports conclusions obtained above from comparison of large regions with different levels of economic development.

7.3 Damage Values and Social Cataclysms

Considering the relation between natural disaster pattern and social-economic development it is pertinent to ask how short a time interval should be for such a relation to be realized. To preliminary discuss this issue we made an attempt to compare changes in disaster pattern in such large countries as Russia and China who have undergone staggering social-economic change in twentieth century and in beginning of twenty-first century.

As to China (area of rather height seismic activity), it is possible to examine changes in vulnerability to seismic impacts occurring during the stormy twentieth century. The first-third of the twentieth century is known to be marked in China by the development followed by a period of intervention and civil war. The period of a

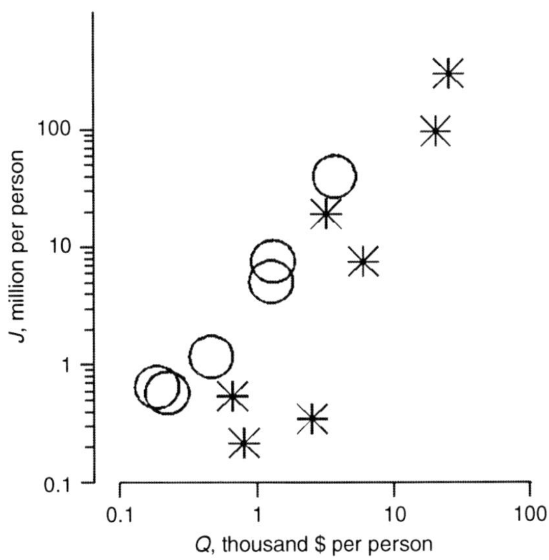

Fig. 7.4 Change of "loss/casualties" ratio J with the annual per capita national product Q based on Table 7.1 (*circles*) and for earthquakes occurring in large cities (*asterisks*)

relatively successful development after the Second World War was followed by the years of the Cultural Revolution. Since the early 1980s China witnessed a stable and fast economic growth. Such a history of development was reflected in the seismic disaster pattern. Figure 7.5 show the death toll during strong earthquakes normalized to earthquake energy and to a current population of China. In accordance with a mean world tendency, there is a tendency of decrease in relative losses from seismic disasters, but it is complicated by a relative increase in vulnerability during the 1970s and early 1980s. Since the mid-1980s there is some growth in a number of events but the norm death toll values tend to decrease. The growth of a number of earthquake disasters can be explained by a more complete record of losses from earthquakes, whereas a decrease in normalized number of the death toll would be reasonably to attribute to a fast economic growth in China.

There was insufficient for statistical analysis number of strong earthquakes in Russia in twentieth century, therefore the database from the Laboratory of the geological risk analyzes of Institute of Geoecology of the Russian Academy of Sciences was used. This database includes data on major natural and natural-technogenic disasters that took place in the present territory of Russia since the X century till present [RUC1]. The database contains information on more than 1,300 extreme events different in nature, loss values, and other characteristics. From this data set 193 mostly hazardous events, that had cause an emergency situation of the

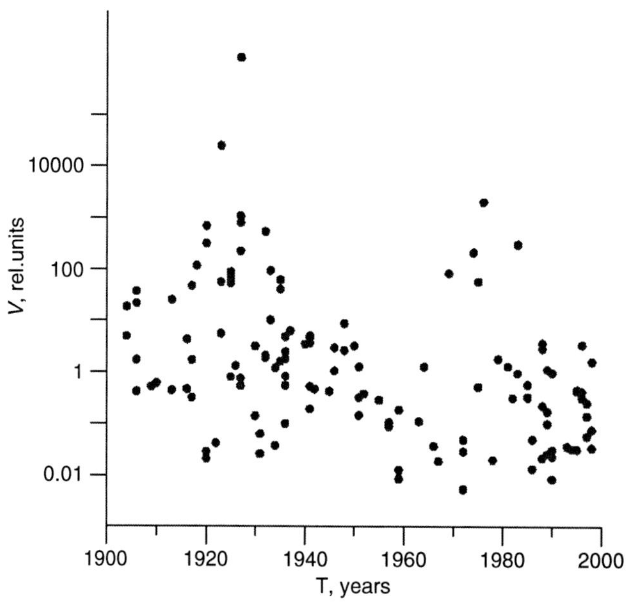

Fig. 7.5 The death toll V (in arbitrary units) normalized on the population of China and earthquake energy. The relative growth of vulnerability of society during the Civil War and intervention, and in the later years of the Cultural Revolution along with a marked decrease in normalized death tolls toward the end of the twentieth century

federal level, have been chosen. According to the database compilers such a selection provides an acceptable completeness and uniformity of data [RUC2].

Figure 7.6 shows an annual numbers of strong disasters (dots) and current 5-years average numbers of disasters (line) occurring since 1900 year. The number of disasters increases very essentially in 1980th that could be connected with an increase in vulnerability of society because of worsening of social-economic situation. Long delay in wages and salaries could not strengthen the labor discipline and work performance rules, that can strongly affect a probability of occurrence and gravity of man-made disasters. Besides, during the crisis there were no resources for carrying out expensive, preventive measures.

Figure 7.6 (may be accidentally) shows also a decrease in mean number of disasters in the mid-1990s, and a new weaker peak during 1998–1999. Hypothetically, it seems possible to associate these peculiar features with economic revival in Russia in the mid-1990s and with the default occurring in 1998 correspondingly. The 1998 default also resulted in delay of salaries, decrease in the labor discipline, and freezing of fulfillment of preventive measures for risk reduction.

Thus, even a rather short-term change in social and economic situation appear can affect the vulnerability of a society from natural and technogenic disasters, and the data for Russia and China appear to support this hypothesis. If we assume, that though extensive preventive engineering measures cannot be accomplished during a short time interval, still some infrastructure, management, and communication

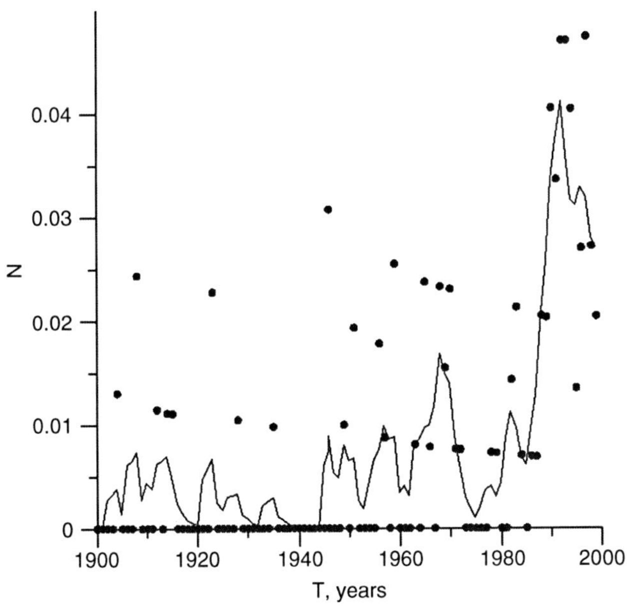

Fig. 7.6 Annual number of major disasters in Russia per million population (*dots*) and averaged values on overlapping 5-year intervals (*line*) (Data from Ragozin et al. [RUC1])

systems might be markedly improved that can also decrease the damage from disasters.

The above presented conclusions about the relation between seismic losses and social-economic conditions agree with results obtained by other authors. A statement that regions with a fairly low level of economic development are less resistant to earthquake impact was done in [So2]. Similar conclusions about the relation between damage amount from natural disasters and a level of social-economic development are presented by Kahn [Ka2]. The author examines relations between death toll from 225 major natural disasters occurring in 1990–2001 and social-economic situation in different countries. As a result, a few empirical relations were revealed. For example, the growth of gross national product from 2,000 to 15,000 dollars per capita per year corresponds meanly to a decrease in a number of casualties by 500 people in year per a 100 million of population. An important factor affecting an average death toll from disasters is also a management structure; countries with more developed democratic institutions turned out to be relatively less vulnerable to disasters.

7.4 The Natural Disasters and the Concept of Sustainable Development

The above mentioned empirical relationship makes doubt in validity of a pessimistic prediction of a catastrophic growth of losses from natural disasters mentioned in the beginning of the chapter. In fact, the growth in number of disasters is mainly caused by the better registration of weak disasters. As to a nonlinear growth of loss values with time, this effect (at least partly) arises from the power-law distribution of loss values from strong disasters.

Having in mind the revealed relations between the seismic disasters and social-economic parameters it seems reasonable to assume a decrease in intensity of the disaster flow with a large number of casualties in the developed countries, and then an extension of this tendency to the developing countries (as their economic and social situation develops and as the rate of population growth slows down). Whereas the tendency of growth of absolute values of material losses will be preserved, a decrease of normalized loss values (in units of GDP) can be expected.

An attention is called to a marked uniformity of normalized values of damage from earthquakes in different countries. There is an impression, that different social communities and mankind as a whole have adapted in the course of historical development to a certain permissible (depending on concrete social-economic conditions) level of losses from natural disasters. It is possible to assume that such a level corresponds to an optimum between the profit from the use of some or other natural resource and losses from natural disasters inherent with the use of this resource (for example, the use of fertile soils in river valleys presupposes flood losses). Thus, in case of damages from natural disasters, the principles of sustainable development turn to be implemented. Such an example of the realization of concept of sustainable development in practice promises that this optimistic concept can be realized also under more complicated conditions. An example of a practical realization of the concept of sustainable development becomes even more important, because the discussion on this issue is dominated by a pessimistic viewpoint, namely, that "despite unprecedented large scale discussion of sustainable development, the world keeps developing along the trajectory of non-sustainable development" [KL, p. 598].

7.5 Main Results

According to a common view the growth in death toll and losses from natural disasters is connected with an increase of population, the development of potentially hazardous industries, and general deterioration of environment. A standard extrapolation of growth of mean values of damage with time makes it possible to predict, that the entire economic gain can be taken up by still increasing losses from natural disasters by the mid-twenty-first century. However, the statistical approach based on mean values is not quite correct in this particular case. The empirical distributions of loss values from natural disasters, as a rule, correspond to the power heavy tail distribution law with the infinite mathematical expectation which makes the using of sample means incorrect.

The application of statistically correct approaches affects the prediction essentially. The catalogs of losses for the case of seismic disasters are the most complete, and it is this case which was examined in more detail. In accordance with a common tendency typical of losses from different kinds of natural disasters, the number of earthquake disasters tends to grow with time, and the number of victims and economic losses from earthquakes reveals a nonlinear growth of cumulative losses with time. But this growth can be explained, basically, not by change of regime of seismic disasters with time, but by the heavy-tailed character of the distribution of the number of victims and economic losses from earthquakes. Evidently also, that because of natural restrictions on the size of the maximum possible death toll and loss values, this nonlinear growth of damage cannot last infinitely. The change in the mode of growth of cumulative losses from the non-linear to the linear pattern was found to occur after a time interval of 30–40 years. Taking into account the non-linear and linear patterns of loss growth with time, we propose a prediction of expected cumulative earthquake losses. The time interval of a valid prediction is restricted by current change of socio-economic conditions. The prediction for the time interval of 50 years is presented.

An examination of the relationship between the earthquake loss values and the changes in social and economic situation in different regions was carried out, and some uniformity of normalized loss level in different regions has emerged. This tendency could be general in nature and correspond to a certain optimum between the advantage gained from the use of natural resources and the losses from disasters connected with the use of these resources. Despite of an increase of absolute losses from disasters, normalized values of losses will tend to decrease with the social-economic development. This behavior is consistent with the concept of sustainable development. Even though the results described here are based mainly on statistics of seismic disasters, the authors believe that the results have a more general character and can be applied to other kinds of natural disasters. The example of the realization of concept of sustainable development in practice promises that this optimistic concept can be realized probably under more complicated conditions also.

Summary and a Review

The present monograph provides a general overview of the state-of-the-art in the statistical analysis of the occurrence of natural disasters and their impact on man. The empirical distributions of disaster losses are peculiar in that they are described by heavy-tailed distributions. The treatment of heavy-tailed distributions using the statistical techniques based on means and variances is limited in scope and can lead to incorrect inferences. It is the development, discussion, and application of correct mathematical approaches which is the main content of the present monograph. We propose a set of statistical methods that makes it possible to make correct analyses of disaster occurrence throughout the practically important range of disaster magnitude, from typical moderate-sized events to rare, extremely large disasters. The case of seismic disasters is treated in the greatest detail because more complete statistical material is available for this case. We describe several novel statistical approaches to the estimation of earthquake hazard and develop a forecast of characteristic cumulative earthquake losses for the future 50 years. We discuss tendencies in the variation of characteristic losses over time for regions with differing levels of socio-economic development. The results provide substantial refinements on the prevailing notions as to relationships between disaster losses and the level of socio-economic development in a country.

The first chapter briefly reviews the state-of-the-art in the study of natural disaster occurrence, viz., we give an account of the diversity and completeness of available factual material, describe the terminology and the main methodological and theoretical concepts. The focus is on empirical distributions which describe observations of various disasters, both the geophysical characteristics of disasters and the associated losses are discussed. It is shown that, in spite of the fact that the empirical distributions of disaster size show a great diversity, they can be described (at least as concerns their leading features) by three type distributions, viz., the normal (Gaussian), the exponential (Boltzmann), and the self-similar power-law distribution (Pareto), or else by a combination of these distributions.

The qualitative differences between the type distributions reflect the essential differences between the processes that generate different kinds of natural disasters. With this essential difference in mind, we suggest dividing all natural disasters into three types: disasters of trend, disasters of extremum, and disasters of breakdown. The quantity which is to discriminate between the disasters is R, the dimensionless ratio of the characteristic value of the damaging factor of a disaster to the background value of an excitation that is similar physically (for example, the ratio of wind velocity in a hurricane to the mean wind velocity). It is shown that the disasters belonging to the different types (i.e., having substantially different values of R) show qualitative differences throughout their main properties, as well as are different in the efficiency achieved with different kinds of loss mitigation measures. The disasters of the first type (trend disasters) are the slowest and the best predictable; loss mitigation in this case is best achieved by preventive measures. The disasters of the third type (breakdown disasters) are the most sudden and the least predictable; the purpose of loss mitigation is best served in this case by insurance systems and response to emergencies.

The classification scheme proposed here is a convenient tool providing orientation among the entire gamut of natural disasters and seems to be useful for the planning of measures to reduce the losses due to all natural and man-induced disasters. In developing the set of measures intended to reduce the losses it seems useful to parameterize disasters of different kinds (floods, hurricanes, etc.) by a parameter set consisting of an intensive parameter R and an extensive parameter L, which describes the loss value due to the disaster of interest, a value that is typical of the region in hand. This parameterization method seems at once very compact and sufficiently informative.

In application to the main content of this monograph – the analysis of the statistical properties of disasters described by heavy-tailed distributions – the proposed classification scheme is convenient in that it characterizes the place of heavy-tailed disasters among the entire diversity of natural and man-induced disasters. It appears from this scheme that it is the disasters of the third type which pose the greatest threat, being as they are the most sudden and causing the greatest damage. These disasters entail the greatest numbers of casualties; it is these disasters which are implied in the word "disaster" in collective consciousness and mass media. The type-creating feature for the disasters of the third type consists in the large values of R and the enormous differences in the range of disaster size, usually several orders of magnitude. The distributions of size for large disasters of the third type are commonly described by power-law distributions, involving the great contribution of single very large events into the total effect (loss) due to all disasters of this kind. This type of distribution is relevant to the physical characteristics of earthquakes, avalanches, the large floods, as well as to the losses caused by large disasters of practically all kinds, both natural and man-induced.

This abnormally high contribution of extremely large individual events into the total effect has several purely mathematical consequences, in particular, the fact that the ordinary statistical characteristics commonly used, the mean and variance, are of limited applicability in the study of disasters of the third type. We remind for comparison purposes that when one has to deal with events described by the normal distribution, the contribution of the largest event into the normalized total effect for N events is of the order of $logN/N$ and vanishes as N increases indefinitely.

This is not the case for heavy-tailed distributions, and the loss due to the greatest event can have in this case the same order of magnitude as the total loss.

The classification of disasters by distribution type is supplemented in the second chapter by a discussion of the conditions in the evolution of individual events that favor the occurrence of the relevant type distribution. The type distributions occur frequently in practice, which is explained by the fact that the conditions giving rise to these distributions are frequent. We investigated the processes giving rise to the occurrence of different type distributions by using simple multiplicative and cumulative models for the generation of individual events. We have constructed schemes that give rise to the different type distributions, thereby developing a mutually consistent and mutually complementary scheme for describing the disaster processes in terms of type distributions of disaster size and in terms of schemes which give rise to the relevant type distribution.

One can see by examining the multiplicative model, which produces power-law distributions, that the general requirement for a power-law distribution to take place is the existence of disequilibrium processes that create positive feedbacks. This conclusion is in our opinion an analogue of the well-known statement [TK] that the presence of (at least one) positive cycle in the dynamics of a system is a necessary condition for the occurrence of a complex multiplicative behavior involving rapid transitions from one state to another.

We have demonstrated that our description of disasters by the type of distribution and by the schemes which give rise to the relevant type distribution is effective. The results derived within the framework of this self-consistent description have shed light on several peculiar features in the occurrence of earthquakes and hurricanes.

In a more complex case where the empirical distribution shows features that are special to both the exponential distribution (for moderate-sized events) and the power-law distribution (for rare, very large events), we demonstrate that a mixture of the multiplicative and the cumulative schemes can be used. The mixed model seems to be useful for monitoring the safety of complex, combined natural-manmade systems. The most important practical application of this model is in situations (rather frequent in practice) in which the return period of large disasters is much shorter than could be expected from extrapolating an empirical distribution of moderate-sized disasters to the region of rare, large events. The mixed model provides a simple, but effective way of describing such a situation.

The assumptions underlying the multiplicative model essentially boil down to admitting the existence of many thermodynamically disequilibrium metastable states capable of avalanche-like energy release. In application to the earth sciences, this assumption corresponds to the concept of M.A. Sadovsky as to an internal activity of the geophysical medium and to the wide occurrence of thermodynamically disequilibrium (metastable) mineral associations in the lithosphere which can turn into stable phases (under certain conditions) in an avalanche-like manner.

We also note that the above approach to the treatment of the generation of power-law distributions is similar, when viewed from the mathematical standpoint, to the treatments developed in ([Go2; Cz; TK] and the references therein).

The third chapter considers nonparametric methods for estimating the probability of large rare events which do not require any assumptions of exact distribution functions. In spite of the fact that these methods are slightly less effective, they are of a specific practical interest. For example, it may be useful in an analysis of regional seismicity to know the distribution of the time T until the next magnitude record in the region. The distribution of T turns out to be close to the log-normal and has a rather heavy tail. For practical purposes therefore, it is better to estimate the quantity $logT$ and its quantiles.

It also seems useful to develop a nonparametric estimate for the probability with which a given magnitude value will be exceeded during a future time interval τ. The results thus obtained allow statements to be made as to the probability of recurrence for events comparable with the largest past event in the region, as well as finding the uncertainty of this estimate. Another practically important application of the probability of exceedance above a given record level is the calculation of dam height for flood prevention. The nonparametric approaches considered in this monograph seem useful in all those cases in which there is no generally accepted method for parameterizing a given empirical distribution. The nonparametric methods are free of the possible error due to unjustified assumptions about distribution law for an empirical distribution under study. The price to be paid for refusing to make a priori assumptions about distribution law is a somewhat lower estimation effectiveness when using nonparametric methods.

Of special interest is the growth of cumulative effects (in particular, the total disaster losses). When in the stationary mode of occurrence, the cumulative effect is commonly a linear function of time. For this reason it is frequently thought that a nonlinear growth of the cumulative effect testifies to a nonstationary of the process under study. This inference is however erroneous in those cases in which the process is described by a power-law distribution with the exponent $\beta < 1$. For such distributions the mathematical expectation is infinite, and accordingly, the sample means tend to grow indefinitely with increasing sample size. The growth can be mistaken as indicating that the process is not stationary. From this, in particular, one concludes to the necessity of a correct confirmation (or refutation) of inferences drawn by a number of authors as to the mean losses due to natural disasters being nonstationary (rapidly increasing).

The fourth chapter uses earthquake loss data to show the existence of two regimes in which the cumulative effect can grow: a nonlinear and a linear. The nonlinear regime of growth for the characteristic cumulative losses is largely related to the existence of a heavy tail in the empirical loss distribution for large (but not extremely large) seismic disasters. The actual nonstationarity in the occurrence of seismic disasters is of secondary importance. The transition from the nonlinear to the linear growth of the cumulative effect is due to the fact that the maximum possible size of events is restricted. In application to seismicity this transition is related to the fact that the maximum loss is limited, because this planet, its population, and the total cost of technology infrastructure are all limited. The model thus developed was used to predict the characteristic cumulative earthquake losses for a time interval of 50 years (during which time the world with its social and technological attributes can be assumed to experience a moderate change). Further refinement on the forecast requires incorporating the dependence of loss on the socio-economic situation. This issue is discussed in the last chapter.

The fifth chapter is concerned with the distribution of earthquake scalar seismic moments. In recent years seismic moment determinations have become much more numerous, and the respective catalogs (the Harvard catalog, in the first place) are much more complete than the catalogs of earthquake losses. The use of this information, which is as complete as can be possible for natural disasters, gave us an opportunity to investigate the behavior of distribution tails in greater detail than was possible for earthquake loss data. The problem of finiteness of the maximum possible effect applies to the case of seismic moments as well. We have studied, not only models involving sharp truncations of the distribution, but also models with an unlimited continuous density but (in contrast to the ordinary Gutenberg-Richter seismic moment – frequency relation) with a finite mean. Several such models have been put forward before by various authors. The estimates they obtained for the tail, however, turned out to be strongly dependent on the model used.

In order to avoid the model-specific errors in determining the behavior of distribution tails we proposed a new parameterization for the tail of a distribution as an extension of the various models suggested by previous workers. The parameterization in question only assumes a finite mean and a monotone distribution density. The key parameter in the parameterization is the size of the characteristic maximum event (earthquake) M_c which separates the range of moderate-sized events described by the ordinary Gutenberg-Richter relation and the range of rare large earthquakes with an unknown distribution that has a finite mean. The approach that analyzes the growth of cumulative seismic moments yields the most robust estimate of M_c and of recurrence period of such events T_c. The application of this approach to data from different regions resulted in regional estimates of M_c. The variability in M_c for different regions has turned out to be very considerable. A correlation was found to exist between the estimates of M_c for different subduction zones and the seismotectonic setting of these zones (subduction rate, plate age, the stress class). That such relationships exist indicates tectonics as an important factor for the seismic moment recurrence relation, thereby providing evidence against the hypothesis that asserts the universality of the frequency-seismic moment relation [K7].

The sixth chapter considers the distribution in the range of the largest events with seismic moment values $M > M_c$. In this case we deal as a rule with the estimation of quantiles that are beyond the data range, i.e., quantiles of level $q > 1 - 1/n$, where n is the sample size. The reliable estimation of quantiles of levels $q > 1 - 1/n$ can be made only with some additional assumptions. Our approach is based on the general limit theorems for extreme values. The results we obtained permit a better understanding of how the tail in a distribution of earthquake size (seismic energy) behaves. As was mentioned above, different authors have tried to fit the distribution of seismic moments by unlimited functions: the gamma function, the exponential function with a fractional index (the stretched exponential distribution), and the Pareto distribution with an exponent higher than unity. Our results indicate that fitting the distribution of earthquake size by a law limited from the right (by a parameter called M_{max}) is more adequate. But the parameter M_{max} can be highly unstable statistically. The probability density tends to its rightmost value M_{max} at a very low level with a tangency of a high degree ("duck beak" shape). This explains the instabilities in the statistical estimates of M_{max} and the rare occurrence of unexpected and consequently especially destructive large earthquakes in the areas believed to be of low seismicity before. In order to describe the tail behavior of the distribution properly we have suggested a new parameter $Q_q(t)$ – the quantile of a given level q of the maximum magnitude in a future time interval T. In contrast to M_{max}, the integral parameter $Q_q(t)$ is stable and robust.

We have modified the proposed method of statistical estimation of tail parameters to deal with continuous seismic moment values, as well as earthquake catalogs with discrete magnitudes.

Chapters 3, 4, 5, and 6 present a set of approaches that allow a correct analysis of disaster (earthquakes) occurrence throughout the whole range of disaster magnitude, and throughout a variety of time spans. Hazard assessment for short time intervals and moderate-sized events is based on the parameters calculated for the ordinary Gutenberg-Richter relation. For longer time intervals when a finite size of the possible maximum event should be taken into account, one should also use the size M_c and return period T_c for events of size M_c. For still longer time intervals it is necessary to calculate the quantiles of rare events larger than M_c using the limit theorems of the extreme value theory. An adequate incorporation of nonlinearities in the frequency-magnitude relation and of specific features in the relation for rare, large events of size greater than M_c is evidently important in connection with earthquake hazard assessment for megacities and critical facilities with very large maintenance times (e.g., for depositories of radioactive waste).

We also emphasize that the methods used here to parameterize the recurrence of various-sized earthquakes are very general and are applicable, not only to problems arising in the assessment of earthquake hazard, but also to all those cases where the recurrence of rare large events is to be estimated. The demonstration of the capability of these new statistical techniques by using earthquake catalogs is solely due to the fact that more extensive and reliable statistical materials are available for this case.

The last, seventh chapter considers relationships between the occurrence of natural disasters and changes in socio-economic conditions. A novel (essentially nonlinear) method is put forward for assessing the expected losses caused by natural disasters based on the incorporation of peculiar properties specific to distributions with heavy tails. The method was applied to seismic disaster data. The result was to identify a quasi-stationarity in the occurrence of earthquake losses, which tend to decrease (when normalized by population number and per capita GNP) during the progressive economic development of different regions. In this connection it has proved feasible to treat the occurrence of disaster losses as an example of possible practical applications inherent in the concept of sustainable development rather than as a nonstationary process of rapid, catastrophically growing losses. The estimate we have derived for earthquake losses is substantially different from the previous forecasts of losses which were usually based on linear regression equations. The forecast incorporating the presence of a heavy tail in empirical loss distributions is much more optimistic. We wish to stress, however, that this forecasts should be exclusively understood as an average tendency, similarly to any other forecast involving random factors and subject to random deviations that might occur in actual conditions. The contribution of such random factors is especially large when the data in hand are described by a heavy-tailed distribution. Thus, in particular, the scatter shown by different random samples of seismic losses will in most cases be substantially higher than the loss changes due to the transition from a nonlinear to a linear growth of characteristic loss.

We note that, in contrast to seismicity, the occurrence of disasters related to climate change and manmade disasters may prove to be much more dependent on time (non-stationarity). Nevertheless, the preliminary analysis of death toll from floods testifies for the evolution of M_c and T_c with time similarly to that of the death toll from earthquakes. In both cases, the death tolls from earthquakes and floods, the value of M_c was found to be lower for more recent times.

It is also of a certain practical interest to try to identify short-term (a few years to a few tens of years) relationships between the recurrence of disaster losses and changes in the socio-economic situation. Some observations point to the fact that even comparatively short-term changes in the socio-economic situation (like the crisis that occurred during the Russian "perestroika") significantly affect the recurrence of losses due to natural disasters. A system that is less stable in the socio-economic respect also shows higher vulnerability when faced with natural and combined natural-manmade disasters. Rapid changes in the social sphere and technological infrastructure occurring in Russia can even be seen from space. For example, changes in nighttime lights clearly detect crises that have affected many industrial centers and the mansion construction boom in the southern and western suburbs of Moscow.

References

[Ab1] Abe K (1982) Magnitude, seismic moment and apparent stress for major deep earthquakes. J Phys Earth 30(4):321–330

[Ab2] Abe K (1995) Modelling of the runup heights of the Hokkaido-Nansei-Oki tsunami of 12 July 1993. PAGEOPH 144:735–745

[An] Ananiev GS (1998) Catastrophic processes in relief generation. Izd-vo MGU, Moskva, 104 pp

[Ba] Bak P (1990) Self-organized criticality. Physica A 163:403–409

[Bar] Barsukov VL (ed) (1981) Sketches of comparative planetology. Nauka, Moscow, 326 pp

[BK] Bird P, Kagan YY (2004) Plate-tectonic analysis of shallow seismicity: apparent boundary width, beta, corner magnitude, coupled lithosphere thickness, and coupling in seven tectonic settings. Bull Seismol Soc Am 94(6):2380–2399

[Bl] Blong R (2003) A review of damage intensity scales. Nat Hazards 29:57–76

[Bo] Borisenkov EP (ed) (1988) Fluctuations of climate during the last millennium. Leningrad, Gidrometeoizdat, 408 pp

[BP] Bender BK, Perkins DM (1993) Treatment of parameter uncertainty and variability for a single seismic hazard map. Earthq Spectra 9(2):165–195

[Br] Brune JN (1968) Seismic moment, seismicity, and rate of slip along major fault zones. J Geophys Res 73:777–784

[BS] Bolshev LN, Smirnov NV (1983) Tables of mathematical statistics. Nauka, GIFML, Moscow, in Russian

[Bu1] Burstein LM (2006) Statistical estimation of parameters of size distribution of oil fields in poorly explored basins. Russ Geol Geophys 47(9):999–1010

[Bu2] Bullen KE (1975) The earth's density. Wiley, New York

[CD] Coles S, Dixon M (1999) Likelihood-based inference for extreme value models. Extremes 2(1):5–23

[CFL] Cosentino P, Ficara V, Luzio D (1977) Truncated exponential frequency-magnitude relationship in the earthquake statistics. Bull Seism Soc Am 67:1615–1623

[Ch] Christopeit N (1994) Estimating parameters of an extreme value distribution by methods of moments. J Stat Plann Inf 41:173–186

[CM] Castellaro S, Mulargia F (2002) What criticality in cellular automata models of earthquakes. Geoph J Int 150:483–493

[Cr] Cramer H (1940) Mathematical methods of statistics. Princeton University Press, Princeton, NY

[Cz] Czechowski Z (2003) The privilege as the cause of power distribution in geophysics. Geophys J Int 154:754–766

[Da] Dargahi-Noubary GR (1983) A procedure for estimation of the upper bound for earthquake magnitudes. Phys Earth Planet Inter 33:91–93

[Dar] Dargahi-Noubary GR (2000) Statistical methods for earthquake hazard assessment and risk analysis. Nova Science Publishers, Huntington, NY

[DES] Dziewonskii AM, Ekstrom G, Salganic MP (1993) Centroid-moment tensor solution for January–March, 1992. Phys Earth Pl Int 77:143–150

[DK] Dacy DC, Kunreuther H (1969) The economics of natural disasters. The Free Press, Toronto, Ontario, p 328

[DS] Davison FC, Scholz CH (1985) Frequency-moment distribution of earthquakes in the Aleutian Arc: a test of the characteristic earthquake model. Bull Seismol Soc Am 75 (5):1349–1361

[DV] Daley DJ, Vere-Jones D (1988) An introduction to the theory of point processes. Springer-Verlag, New York

[EKM] Embrechts P, Kluppelberg C, Mikosch T (1997) Modelling extremal events. Berlin-Heidelberg, Springer-Verlag, 645 pp

[EL] Epstein BC, Lomnitz C (1966) A model for the occurrence of large earthquakes. Nature 211:954–956

[EMB] El-Sabh ML, Murty TS, Briand J-M (1988) Negative storms surges on Canada's east coast. In: El-Sabh MI, Murty TS (eds) Natural and man-made hazards. Dordrecht, Holland, pp 395–410

[Ef] Efron B (1988) Unconventional methods of multivariate statistical analysis. Finansy i Statistika, Moscow, 264 pp, in Russian

[En] Engle RF (1982) Autoregressive conditional heteroscedastic models with estimates of variance of United Kingdom inflation. Econometrica 50:987–1007

[Fe] Feller W (1966) An introduction to probability theory and its applications, vol II. Wiley, New York

[Fr] Frisch U (1995) Turbulence, the legacy of A.N. Kolmogorov. Cambridge University Press, Cambridge

[Fu] Fujita T (1981) Tornadoes and downbursts in the context of generalized planetary scales. J Atmos Sci 38:1511–1534

[Ga] Galambos J (1978) The asymptotic theory of extreme order statistics. Wiley, New York

[GD] Georgio PN, Davenport AG (1988) Estimation of the wind hazard in tropical cyclone regions. In: El-Sabh ML, Murty TS (eds) Natural and man-made hazards. Dordrecht, Holland, pp 709–726

[GGG] Galadini F, Galli P, Giraudi C (1997) Geological investigations of Italian earthquakes: new paleoseismic data from the Fucino plain (central Italy). J Geophys 24:87–103

[Gi] Van Gigch JP (1978) Applied general systems theory, 2nd edn. Harper, New York

[GMMY] Grachev AF, Magnitskii VA, Mukhamediev SA, Yunga SL (1996) Determination of possible maximum magnitudes of earthquakes in the East European Platform. Physica Zemli 7:3–20, in Russian

[Go1] Golitsyn GS (1997) Statistics and energetic of tropical cyclones. Proc Russ Acad Sci 354(4):535–538, in Russian

[Go2] Golitsyn GS (2001) The place of the Gutemberg-Richter law among other statistical laws of nature. Comput Seismol 32:138–161

[Go3] Golitsyn GS (2005) Size of cyclones and statistical properties of their ensemble. Proc Russ Acad Sci 401(1):72–74

[GP] Godano C, Pingue F (2000) Is the seismic moment-frequency relation universal? Geophys J Int 142:193–198

[GPRY] Golitsyn GS, Pisarenko VV, Rodkin MV, Yaroshevich MI (1999) Statistical characteristics of parameters of tropical ciclons and problem of estimation of risk. Proc Russ Acad Sci Phys Atm Ocean 35:734–741

[GR] Gutenberg B, Richter C (1954) Seismicity of the earth, 2nd edn. Princeton University Press, Princeton, NJ 79:645–654

[Gu1] Gumbel EI (1958) Statistics of extremes. Columbia University Press, New York

[Gu2] Gushchenko II (1979) Eruptions of volcanoes of the world: a catalog. Nauka, Academic of Science USSR Far East Science Center, Moscow, 474 p

[Gu3] Gusev AA (1989) Multisparity fault model and the nature of short-period subsources. Pure Appl Geophys 130:635–660

[Ha] de Haan L (1990) Fighting the atch-enemy with mathematics, Statist. Neerlandica 44:45–68

[Har] Hardle W (1989) Applied nonparametric regression. Cambridge University Press, Cambridge

[Hi] Hill BM (1975) A simple general approach to inference about the tail of a distribution. Ann Statist 3:1163–1174

[HK] Hogg RV, Klugman SA (1984) Loss distributions. Wiley, New York, 235 pp

[Ho] Holland GJ (1982) An analytical model of the wind and pressure profile in hurricanes. Month Weath Rev 108:1212–1218

[HS] Ho C-H, Smith EI (1997) Volcanic hazard assessment incorporating expert knowledge: application to the Yucca Mountain region, Nevada, USA. Math Geol 29:615–627

[Ja] Jarrard RD (1986) Relations among subduction parameters. Rev Geophys 24(2): 217–284

[K1] Kagan YY (1991) Seismic moment distribution. Geophys J Int 106:123–134

[K2] Kagan YY (1994) Observational evidence for earthquakes as a nonlinear dynamic process. Physica D 77:160–192

[K3] Kagan YY (1996) Comment on "The Gutenberg-Richter or characteristic earthquake distribution, which is it?" by Steven G. Wesnousky. Bull Seism Soc Am 86:274–285

[K4] Kagan YY (1997a) Seismic moment-frequency relation for shallow earthquakes: Regional comparison. J Geophys Res 102:2835–2852

[K5] Kagan YY (1997b) Earthquake size distribution and earthquake insurance. Commun Statist Stachastic Models 13(4):775–797

[K6] Kagan YY (1997c) Are earthquakes predictable? Geophys J Int 131(3):505–525

[K7] Kagan YY (1999) Universality of the seismic moment-frequency relation. Pure Appl Geophys 1:537–573

[K8] Kagan YY (2002a) Seismic moment distribution revisited: I Statistical results. Geophys J Int 148:520–541

[K9] Kagan YY (2002b) Seismic moment distribution revisited: II Moment conservation principle. Geophys J Int 149:731–754

[Ka1] Kagan YY (1993) Statistics of characteristic earthquakes. Bull Seismol Soc Am 83 (1):7–24

[Ka2] Kahn EM (2003) The death toll from natural disasters: the role of income, geography and institutions http://ssrn.com/abstract=391741 or doi: 10.2139/ssm.391741 pp, University of California, Los Angeles

[Kas] Kasahara K (1981) Earthquake mechanics. Cambridge University Press, Cambridge, Tokyo

[KDS] Kontorovich AE, Demin VI, Strakhov IA (1985) The principles of discovering oil and gas fields of various reserves in oil and gas basins. Geol Geophys 11:3–15

[Ke] Kesten H (1973) Random difference equations and renewal theory for products of random matrices. Acta Mathemat 131:207–248

[KG] Kijko A, Graham G (1998) Parametric-historic procedure for probabilistic seismic hazard analysis, Part I, Estimation of Maximum Regional Magnitude Mmax. Pure Appl Geophys 152:413–442

[Ki] Kijko A (2004) Estimation of the maximum earthquake magnitude, M*max*. Pure Appl Geophys 161:1–27

[KK] Knopoff L, Kagan Y (1977) Analysis of the Extremes as Applied to Earthquake Problems. J Geophys Res 82:5647–5657

[KKK] Knopoff L, Kagan Y, Knopoff R (1982) b-Values for foreshocks and aftershocks in real and simulated earthquake sequences. Bull Seism Soc Am 72(5):1663–1675

[KL] Kondratiev KY, Losev KS (2002) Illusions and reality of the sustainable development strategy. Vestnik RAN 72(7):592, in Russian

[KLG] Kijko A, Lasocki S, Graham G (2001) Non-parametric seismic hazard in mines. Pure Appl Geophys 158:1655–1675

[KN] Kronrod TL, Nekrasova AK (1996) Seismic risk in largest cities of the world: assessment verification. Comput Seismol 28:225, in Russian

[Ko] Kolmogorov AN (1941a) Local structure of turbulence in an incompressible viscous liquid for very large Raynolds numbers. Doklady AN USSR 30:9–13, in Russian

[Kol] Kolmogorov AN (1941b) On log-normal distribution law of particle sizes in fragmentation process. Doklady AN USSR 31:99–101

[KPR1] Kuznetsov IV, Pisarenko VF, Rodkin MV (1997) Methods for calculating losses due to disasters of different types. Ekonomika i Matematicheskie Metody 33(4): 39–50

[KPR2] Kuznetsov IV, Pisarenko VF, Rodkin MV (1998) On the classification of disasters: parameterization of impact and loss. Geoekologiya 1:16–29

[Ks1] Kijko A, Sellevol MA (1989) Estimation of earthquake hazard parameters from incomplete data files. Part I, Utilization of extreme and complete catalogues with different threshold magnitudes. Bull Seism Soc Am 79:645–654

[Ks2] Kijko A, Sellevol MA (1992) Estimation of earthquake hazard parameters from incomplete data files. Part II, Incorporation of magnitude heterogeneity. Bull Seism Soc Am 82:120–134

[Ks3] Knopoff L, Sornette D (1995) Earthquake death tolls. J Phys I (France) 5:1681–1688

[Ks4] Kagan YY, Schoenberg F (2001) Estimation of the upper cutoff parameter for the tapered distribution. J Appl Probab 38A:901–918

[Ku] Kukal Z (1985) Natural disasters. Znanie, Moscow, 240 pp, in Russian

[Lo] Lomnitz C (1994) Fundamentals of earthquake prediction. Willey, New York, 220 p

[LS] Laherrere J, Sornette D (1998) Streched exponential distributions in nature and economy: "fat tails" with characteristic scales. Eur Phys J 2:525–539

[LTPK] Lyubushin AA, Tsapanos TM, Pisarenko VF, Koravos GC (2002) Seismic hazard for selected sites in Greece: a Bayesian estimate of seismic peak ground acceleration. Nat Hazards 25:83–98

[M] Main Y (2000) Apparent breaks in scaling in earthquake cumulative frequency-magnitude distribution: fact or artifact? Bull Seism Soc Am 90:86–97

[MIMR] Main Y, Irving D, Musson R, Reading A (1999) Constraints on frequency-magnitude relation and maximum magnitudes in the UK from observed seismicity and glacio-isostatic recovery rates. Geophys J Int 137:535–550

[Ma] Mandelbrot B (1983) The fractal geometry of nature. Freeman, New York, p. 460

[MD] Molchan GM, Dmitrieva OE (1991) Aftershock identification: a review and new approaches. In: Contemporary methods for seismological data processing (Vychislitel'naya seismologiya 24). Nauka, Moscow, pp 19–49, in Russian

[MKDN] Molchan G, Kronrod T, Dmitrieva O, Nekrasova A (1996) Seismic risk oriented multiscale seismicity model: Italy. Vychislitelnaya seismologiya 28:193–224, in Russian

[MKP] Molchan G, Kronrod T, Panza GF (1997) Multi-scale seismicity model for seismic risk. Bull Seis Soc Am 87:1220–1229

[MM] Malamud BD, Morein G et al. (1998) Forest fires: an example of self-organizaed critical behavior. Science 281:1840–1841

[MMS] Mikumo T, Miyatake T, Santoyo A (1998) Dynamic rupture of asperities and stress change during a sequence of large interplate earthquakes in the Mexican subduction zone. Bull Seismol Soc Am 88(3):686–702

[Mo] Molchan GM (1996) Scaling exponents and multifractal dimensions for independent random cascades. Commun Math Phys 179:681–702

[MP] Molchan GM, Podgaetskaya VM (1973) The parameters of global seismicity. In: Computational and statistical methods for interpretation of seismic data (Vychislitelnaya seismologiya 6). Nauka, Moscow, pp 44–66, in Russian

[MRT] Morozov VN, Rodkin MV, Tatarinov VN (2001) On the problem of geodynamic safety of object nuclear-fuel circle. Geoecology 3:125–137

[MS] Mantegna RN, Stanley HE (2000) An introduction to econophysics: correlations and complexity in finance. Cambridge University Press, Cambridge, 148 pp

[M1] Munich Re (1988) World map of natural hazards. Munich, Germany

[M2] Murty TS (1977) Seismic sea waves, tsunamis. Bulletin number 198, Fisheries Research Board of Canada, Ottawa

[NG] Nelson JB, Ganze RA (1980) Significant earthquakes 1900–1979: NGSDC-NOAA. World Data Center A, Boulder, 20 p

[Ni] Nikonov AA (1994) On Great Caucasus strong earthquakes in the A.D. first millennium: a revision of the initial data and catalogue. Physica Zemli 7–8:107–112, in Russian

[NMGPFN] Nurmagambetov A, Mikhailova NN, Golinsky GL, Plotnikova LM, Frolova AG, Negmatullaev SK (2000) Comparative seismic hazard estimation for the capitals of the countries of Central Asia. J Earthq Prediction Res 8:496–505

[O1] Osipov VI (1995) Natural disasters in the focus of scientists. Vestnik RAN 65(6):483–495, in Russian

[O2] Osipov VI (2001) Natural disasters at the eve of the 21th century. Geoekologiya 4:293–309, in Russian

[O3] Osipov VI (2002) Natural hazard control. Vestnik RAN 8:678–686, in Russian

[OK] Ogata Y, Katsura K (1993) Analysis of temporal and special heterogeneity of magnitude frequency distribution inferred from earthquake catalogues. Geophys J Int 113:727–738

[P1] Pisarenko VF (1989) On the law of earthquake recurrence. In: Discrete properties of the geophysical medium. Nauka, Moscow, pp 47–60, in Russian

[P2] Pisarenko VF (1991) Statistical estimation of the maximum possible earthquake. Fizika Zemli 9:38–46, in Russian

[P3] Pisarenko VF (1995) On the best statistical estimator of the maximum possible earthquake magnitude. Dokl RAN 344(2):237–239, in Russian

[P4] Pisarenko VF (1998) Nonlinear growth of cumulative flood losses with time. Hydrol Process 12:461–470

[PL] Pisarenko VF, Lysenko VB (1997) The probability distribution for the maximum earthquake that can occur during a fixed time interval. Fizika Zemli 6:15–23, in Russian

[Pl] Plate-Tectonic Map of the Circum-Pacific Region (1981) Scale 1:10 000 000. Circum-Pacific Council for energy and mineral resources, the American Association of Petroleum Geologists, Tulsa, Oklahoma, USA, 1981

[PLLG] Pisarenko VF, Lyubushin AA, Lysenko VB, Golubeva TV (1996) Statistical estimation of seismic hazard parameters: maximum possible magnitude and related parameters. Bull Seismol Soc Am 86(3):691–700

[Po] Porfiriev BN (1989) Emergency management. Znanie 5:48

[PR] Pelinovsky EN, Ryabov IA (2000) Distribution functions of tsunami run-un heights (on data of International Surveys 1992–1998). Oceanology 40(5):645–652

[PS1] Pisarenko VF, Sornette D (2003) Characterization of the frequency of extreme earthquake events by the generalized pareto distribution. Pure Appl Geophys 160:2343–2364

[PS2] Pisarenko VF, Sornette D (2004) Statistical detection and characterization of a deviation from the Gutenberg-Richter distribution above magnitude 8. Pure Appl Geophys 161:839–864

[PSS] Pacheco JF, Scholz C, Sykes L (1992) Changes in frequency-size relationship from small to large earthquakes. Nature 355:71–73

[PSSR] Pisarenko VF, Sornette A, Sornette D, Rodkin MV (2008) New approach to the characterization of Mmax and of the tail of the distribution of earthquake magnitudes. Pure Appl Geophys 165:1–42

[R1] Rodkin MV (2002) Model of a seismic regime in a medium with a power-law strength distribution. Phys Solid Earth 38(2):149–155

[R2] Rodkin MV (2003) Classification of catastrophic impacts due to natural hazards. In: Natural hazards in Russia, vol 6, Evaluation and control of natural risks. Izd-vo Kruk, Moscow, pp 85–102

[R3] Rodkin MV (2005a) Linear and nonlinear increase regimes of the cumulative seismic moment. Phys Solid Earth 41(2):95–103

[R4] Rodkin MV (2005b) The model of synergetic effect development upon severe disasters. Geoecology 1:81–87, in Russian

[Ra] Rautian TG (1991) The seismically active medium and earthquake sources. In: Modeling and field studies of earthquake sources. Nauka, Moscow, pp 35–48

[RGL] Rodkin MV, Gvishiani AD, Labuntsova LM (2008) Models of generation of power laws of distribution in the processes of seismicity and in formation of oil fields and ore deposits. Russ J Earth Sci 10(5), June

[RK1] Ruff L, Kanamori H (1980) Seismicity and the subduction process. Phys Earth Pl Int 23:240–252

[RK2] Rautian TG, Khalturin VI (1991) Focal earthquake spectra. In: Earthquakes and processes of their preparation. Nauka, Moscow, pp 82–93, in Russian

[RN] Reisner GI, Nikonov AA (1996) Reappraisal of seismic potential of the Caucasus region. Physica Zemli 8:3–12

[RP1] Rodkin MV, Pisarenko VF (2000) Economic damage and casualties from earthquakes: Statistical analysis, Moscow, GEOS. Comput Seismol 31:242–272, in Russian

[RP2] Rodkin MV, Pisarenko VF (2003) Earthquake loss. Moscow, GEOS, Comput Seismol 34:205–209, in Russian

[RP3] Rodkin MV, Pisarenko VF (2004) Earthquake losses and casualties: a statistical analysis. In: Chowdhury DK (ed) Computational seismology and geodynamics, vol 6, American Geophysics Union. Washington, DC, pp 85–102

[RP4] Rodkin MV, Pisarenko VF (2006) Extreme earthquake disasters – verification of the method of parameterization of the character of distribution of the rare major events, Ch 08. Adv Geosci 1:75–89

[RP5] Rodkin MV, Pisarenko VF (2008) Damage from natural disasters: fast growth of losses or stable ratio? Russ J Earth Sci 10(1), ES1004, doi:10.2205/2007ES000267

[RR1] Romanovicz B, Rundle JB (1993) On scaling relation for large earthquakes. Bull Seismol Soc Am 83(4):1294–1297

[RR3] Rodier JA, Roche M (1984) World catalogue of maximum observed floods. International Association of Science Publications No 143, Wallingford, CT

[RRB] Rogojin EA, Reisner GI, Besstrashnov VM (2002) Seismotectonic situation of Sakhalin Island. Physica Zemli 3:35–44 in Russian

[RS1] Rodkin MV, Shebalin NV (1993a) The regime of natural disasters. Priroda 6:68–73, in Russian

[RS2] Rodkin MV, Shebalin NV (1993b) Problems in disaster measurement. Izv RAN ser Geogr 5:106–116, in Russian

[RUC1] Ragozin AL, Uvarova AV, Chumachenko SA (2003) Supplement catalogue of major nonantropogenic disasters in the territory of Russia over 1000 years. In: Prirodnye Opasnosti Rossii, vol. 6, Otsenka i Upravlenie Prirodnymi Riskami. KRUK, Moscow, p 12, in Russian

[RUC2] Ragozin AL, Uvarova AV, Chumachenko SA (2004) First results of the dynamics analysis of major disasters in the territory of Russia. In: Sergeevskie Chtenia, Issue 6,

Inzhenernaya Geologia i Okhrana Geologicheskoi Sredy. GEOS, Moscow, p 185, in Russian

[Ry] Ryle DM (1998) Forecasting sizes and repose times of future extreme volcanic events. Geology 26:367–370

[Sa] Sadovsky MA (ed) (1989) Discrete properties of the geophysical medium. Nauka, Moscow

[Saf] Safety Problems for Russia (1995–1999), Journal of the VNII GOChS MChS, publications of 1995–1999

[SAK] Shakhramanyan MA, Akimov VA, Kozlov KA (1998) Assessment of Russia's safety from natural and man-induced hazards, theory and practice. VNII GOChS, Moscow, 218 pp

[SBP] Sadovsky MA, Bolkhovitinov LG, Pisarenko VF (1987) Deformation of the geophysical medium and the seismic process. Nauka, Moscow, in Russian

[Sc] Scholz CH (1991) Earthquakes and faulting: self-organized critical phenomena with a characteristic dimension. In: Riste T, Sherrington D (eds) Spontaneous formation of spase-time structures and criticallity. Kluwer, Netherlands, pp 41–56

[SGK] Soloviev SL, Go CN, Kim KS (1992) Catalog of tsunamis in the Pacific, 1969–1982. Academy of Sciences of the USSR, Moscow, 208 p

[SGZLB] Silva PG, Goy JL, Zazo C, Lario J, Barbaji T (1997) Paleoseismin indications along "aseismic" fault in the Guadalentin depression (SE Spain). J Geophys 24 (1):105–115

[She] Shebalin NV (1985) Patterns in natural disasters. Znanie, Moscow, 48 pp

[Shi] Shiryaev AN (1998) Principles of stochastic financial mathematics, vol 1. Facts Models, FAZIS, Moscow, 489 pp

[Shn] Shnirman MG (1987) A dynamic multiscale model of fault generation, numerical modeling and analysis of geophysical processes. Nauka, Moscow (Computational Seismology, Issue 20), pp 87–94, in Russian

[SKKV] Sornette D, Knopoff L, Kagan YY, Vanneste C (1996) Rank-ordering statistics of extreme events: application to the distribution of large earthquakes. J Geophys Res 101(6):13883–13893

[So1] Sobolev GA (1993) Principles of earthquake prediction. Nauka, Moscow, pp 313, in Russian

[So2] Sobolev GA (ed) (1997) Assesment of seismic hazard and seismic risk. FTzNTP RP "Bezopasnost", Moscow, 52 pp

[Sol] Soloviev SL (1970) Recurrence of tsunamis in the Pacific. In: Adams WM (ed) Tsunamis in the Pacific Ocean. Honolulu, Hawaii, pp 149–163

[Sor] Sornette D (2000) Critical phenomena in natural sciences. Springer, Berlin, 434

[SS1] Sornette A, Sornette D (1989) Self-organized criticality and earthquakes. Europhys Lett 9:197–202

[SS2] Simkin T, Siebert L (1994) Volcanoes of the world. Geoscence Press, Tucson

[SS3] Shapoval A, Shnirman M (2001) Prediction in a Two-polarity Model of Sandpile Type Comp Seism 32:225–236, in Russian

[St] Stephens MA (1974) EDF statistics for goodness of fit and some comparisons. J Am Statist Ass 68(347):730–737

[Str] Streltsov MI (2005) The May 27 (28), 1995 Neftegorsk earthquake on Sakhalin island. "Yanus-K", Moscow, 178 pp

[SY] Sorochinskii MA, Yushenko GP (1979) The occurrence of major snow storms in the Soviet Union. In: The analysis and prediction of especially hazardous weather occurrences. Gidrometeoizdat, Leningrad, pp 99–110

[TK] Thomas R, Kaufman M (2001) Multistationarity, the basis of cell differentiation and memory. II. Logical analysis of regulatory networks in terms of feedback circuits. Chaos 11:180–195

[TLP] Tsapanos TM, Lyubushin AA, Pisarenko VF (2001) Application of a Bayesian Approach for estimation of seismic hazard parameters in some regions of Circum-Pacific Belt. Pure Appl Geophys 158:859–875

[TM] Turcotte DL, Malamud BD (2004) Landslides forest fires, and earthquakes: examples of self-organized critical behavior. Physica A Statist Mech Applications 340(4):580–589

[Tu] Turcotte DL (1997) Fractals and chaos in geology and geophysics. Cambridge University Press, Cambridge, 398 pp

[U] Utsu T (1999) Representation and analysis of the earthquake size distribution: a historical review and some new applications. Pure Appl Geophys 155:509–535

[UPM] Ulomov VI, Polyakova TP, Medvedeva NS (2002) On the long-term prediction of strong earthquakes in Central Asia and the Black Sea-Caspian Region. Izvestiya, Phys Solid Earth 38(4):276–290

[UP] Uski M, Pelkonen E (2006) Earthquakes in northern Europe. Report R-232, Inst Seis Univ Helsinki

[Ve] Vere-Jones D (1990) Probability and information gain for earthquake forecasting. Comput Seismol Geodyn 5:104–114

[VVM] Vladimirov VA, Vorobiev JL, Malinetskij GG (2000) Management of risk. Risk, sustainable development, synergetics. Moscow, Nauka, 432 pp, in Russian

[VMGAO] Vikulin AV, Melekestsev IV, Gusyakov VK, Akmanova DR, Osipova NA (2009) Complex (catastrophic volcanic + strongest seismic events) electronic data base as a base of geodynamic paradigm (for Pacific region). In: Problems of complex geophysical monitoring of the Far East of Russia. Abstracts of the conference, Petropavlovsk-Kamchatskii, p 13

[VMO] Vorobiev YL, Malinetsky GG, Osipov VI (2000) Disasters and Society. Kontakt-Kultura, Moscow, 332 pp

[W] Ward SN (1997) More on Mmax. Bull Seismol Soc Am 87(5):1199–1208

[Wa] Wassef AM (1991) Relative impact on human life of various types of natural disasters over the period 1947–September 1991. Report for IUGG Committee for Problems of the International Dacade for Natural Disasters Reduction, Center for Surveying Science, Ontario, Canada

[We] Wesnousky SG (1994) The Gutenberg-Richter or characteristic earthquake distribution, which is it? Bull Seism Soc Am 84:1940–1959

[Wo] Woo G (1999) The mathematics of natural catastrophes. Imperial College Press, London, 292 pp

[Wu] Wu ZL (2000) Frequency-size distribution of global seismicity seen from broad-band radiated energy. Geophys J Int 142:59–66

[Yo] Yokoi Y, Carr JR, Watters RJ (1995) Fractal character of landslides. Env Engin Geosci 1:75–81

[Z1] Zolotarev VM (1983) One-dimensional stable distributions. Nauka, Moscow, 304 pp

[Z2] Zolotarev VM (1984) Stable Laws and their applications. In: The new in life, science, and technology. Mathematics and cybernetics, no. 11. Znanie, Moscow, 64 pp

[ZS] Zolotarev VM, Strunin BM (1971) On the distribution of internal stresses under random location of point defects. Fizika Tverdogo tela 13:594–596

Index

LaVergne, TN USA
06 August 2010
192351LV00002B/13/P

9 789048 191703